BrightRED Study Guide

Curriculum for Excellence

N5

APPLICATIONS OF MATHEMATICS

Brian J Logan

First published in 2017 by:
Bright Red Publishing Ltd
1 Torphichen Street
Edinburgh
EH3 8HX

A CIP record for this book is available from the British Library.

ISBN 978-1-906736-78-1

With thanks to: PDQ Digital Media Solutions Ltd. (layout), Anne Horscroft (editorial)

Cover design and series book design by Caleb Rutherford – e i d e t i c.

Acknowledgements
Every effort has been made to seek all copyright-holders. If any have been overlooked, then Bright Red Publishing will be delighted to make the necessary arrangements.

Permission has been sought from all relevant copyright holders and Bright Red Publishing are grateful for the use of the following:

Several diagrams drawn by Leah McDowell, ELEM design; Images licensed by Ingram Image (pp 8, 11, 21, 29, 49, 77, 81, 87, 98, 102, 104 & 108); eskaylim/iStock.com (p 6); nolexa/iStock.com (p 7); AdrianHancu/iStock.com (p 9); monkeybusinessimages/iStock.com (p 10); Klubovy/iStock.com (p 14); happymoron/iStock.com (p 16); nathan4847/iStock.com (p 18); robtek/iStock.com (p 19); anyaberkut/iStock.com (p 20); Rawpixel/iStock.com (p 23); mozcann/iStock.com (p 26); Caleb Rutherford, e i d e t i c (p 27); 06/iStock.com (p 29); Verdateo/Shutterstock.com (p 37); Caleb Rutherford, e i d e t i c (p 40); Trepalio/iStock.com (p 40); Aynur_sib/iStock.com (p 42); German-skydiver/iStock.com (p 43); Microstock Man/Shutterstock.com (p 46); Caleb Rutherford, e i d e t i c (pp 46–47); HSNPhotography/iStock.com (p 48); mikafotostok/iStock.com (p 50); lisafx/iStock.com (p 51); Alexander Kachkaev (CC BY 2.0)[1] (p 53); miskolin/iStock.com (p 54); Raa_Sz/freeimages.com (p 55); mactrunk/iStock.com (p 60); LindaJoHeilman/iStock.com (p 67); Alistair Michael Thomas/Shutterstock.com (p 70); cobrasoft/freeimages.com (p 72); standby/iStock.com (p 73); ARENA Creative/Shutterstock.com (p 74); Skvoor/iStock.com (p 75); Multiart/iStock.com (p 76); vichie81/Shutterstock.com (p 78); Adisa/Shutterstock.com (p 80); AntonioMP/iStock.com (p 83); Hogweard (CC BY-SA 3.0)[2] (p 84); NordNordWest (CC-BY-SA-3.0-DE)[3] (p 85); Xhandler (CC BY-SA 3.0)[2] (p 85); Steve Etherington/LAT Photographic (CC BY-SA 2.0)[1] (p 86); Joe de Sousa (CC0 1.0)[4] (p 88); Ensup/iStock.com (p 89); fotokostic/iStock.com (p 91); Caleb Rutherford, e i d e t i c (p 94); JohnFScott/iStock.com (p 95); Calmac Ferries Ltd (p 95); 1000 Words/Shutterstock.com (p 96); 4FR/iStock.com (pp 98–99); Andrea Kratzenberg/freeimages.com (p 100); Jag_cz/iStock.com (p 105).

[1] (CC BY 2.0) http://creativecommons.org/licenses/by/2.0/
[2] (CC BY-SA 3.0) https://creativecommons.org/licenses/by-sa/3.0/
[3] (CC-BY-SA-3.0-DE) https://creativecommons.org/licenses/by-sa/3.0/de/
[4] (CC0 1.0) https://creativecommons.org/publicdomain/zero/1.0/

Printed and bound in the UK by Ashford Colour Press Ltd.

CONTENTS

INTRODUCING NATIONAL 5 APPLICATIONS OF MATHEMATICS

Welcome to this study guide for National 5 Applications of Mathematics. The fact that you are reading this guide shows that you care about your performance in this course and proves that you are determined to do well and gain the required qualification.

To be successful in National 5 Applications of Mathematics, you will have to prepare properly. You must attend lessons, study the course, practise key examples and ask for advice about areas of concern. This guide is designed to help you achieve your aims, but it is not enough on its own. You can ask your teacher or lecturer for advice, discuss the course with friends and possibly attend study classes.

HOW TO USE THIS BOOK

The book consists of three chapters covering the National 5 course. These chapters cover the three units of the course: **Managing finance and statistics; Geometry and measures**; and **Numeracy**. There are 46 two-page spreads covering the major aspects of National 5 Applications of Mathematics. It is possible that you will be taught topics in a different order from that given in this guide, although this is the order in which the units appear on the Scottish Qualifications Authority (SQA) website. You may, for example, decide that it is best to start with the part on Numeracy as this contains many important themes that run through the first two parts, such as whole numbers, rounding, decimals and fractions. It is up to you to decide the order of study. You can organise your study to fit in with your class lessons and then use the guide to help prepare for your exam.

Each spread contains the key elements from each topic. There is a **Don't forget** section that includes vital areas to focus on within each spread. Each spread contains advice, formulae, diagrams and examples of the standard you can expect in the exam, together with solutions and hints. Each spread ends with a section called **Things to do and think about**, which contains examples for you to practise. The solutions for all of these examples can be found on pages 109–111.

Another innovation in this guide is an online source of examples and solutions covering the entire syllabus. Solutions are given for all the examples. This should provide a useful resource as you study. There are also suggested video links in each spread. These include YouTube clips of teachers in action, visual guides, online calculators and even the occasional song. As you work through the book, try the examples before you read the solutions.

THE EXAM

You will sit an exam consisting of 2 question papers. The exam is worth a total of 110 marks. The exam usually takes place in the early part of May.

Paper 1 will be a non-calculator paper worth 45 marks. Paper 1 will last 1 hour and 5 minutes.

In paper 2, you will be allowed to use a calculator. Paper 2 is worth 65 marks. Paper 2 will last 2 hours.

CASE STUDIES

Paper 2 of the exam will contain longer questions which focus on a particular theme, for example, going on holiday. These questions will test skills from different units. The study guide includes four case studies with detailed solutions, as well as a number of case studies for you to practise your skills.

FORMULAE

You will have access to some important formulae in your assessments.
Any formulae not on the following list will have to be memorised:

Circumference of a circle: $C = \pi d$

Area of a circle: $A = \pi r^2$

The Theorem of Pythagoras: $a^2 + b^2 = c^2$

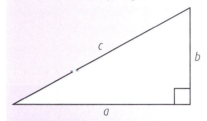

Volume of a cylinder: $V = \pi r^2 h$

Volume of a prism: $V = Ah$

Volume of a cone: $V = \frac{1}{3}\pi r^2 h$

Volume of a sphere: $V = \frac{4}{3}\pi r^3$

Standard deviation: $s = \sqrt{\dfrac{\Sigma(x - \bar{x})^2}{n - 1}} = \sqrt{\dfrac{\Sigma x^2 - (\Sigma x)^2/n}{n - 1}}$, where n is the sample size

Gradient: $\text{Gradient} = \dfrac{\text{vertical height}}{\text{horizontal distance}}$

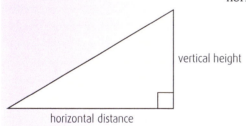

SQA WEBSITE

The SQA website may help with your study as it gives information about the course as well as practice material. Head to www.sqa.org.uk, choose Applications of Mathematics from the subject list and search for National 5 Applications of Mathematics.

Best wishes for all your studying this session!

PLANNING A BUDGET

WHAT IS A BUDGET?

In the novel *David Copperfield* by the great English author Charles Dickens, the character Mr Micawber uttered the following well-known quotation:

> *Annual income twenty pounds, annual expenditure nineteen pounds nineteen shillings and sixpence, result happiness. Annual income twenty pounds, annual expenditure twenty pounds and sixpence, result misery.*

You do not have to understand pounds, shillings and pence to realise that happiness occurs when you spend (expenditure) less than you can afford (income). A **budget** is a way of planning income and expenditure to ensure that you are only spending what you can afford. When this happens, you will have money left over and there will be a **surplus**. When you spend more than you can afford, there will be a **deficit** and you will be in debt.

HOUSEHOLD BUDGETING

In all households, families have to organise their budget so that their income is more than their expenditure, leading to a surplus. Income could come from employment, pensions, tax credits, benefits, interest and **dividends**. Expenditure could arise through household bills for gas, electricity, telephone, council tax and insurance; through expenses for a **mortgage** or rent; on cars and transport; not to mention food, clothes, furniture, entertainment and holidays.

DON'T FORGET

When income > expenditure, there is a surplus; when income < expenditure, there is a deficit.

ONLINE

Investigate a budget calculator at www.brightredbooks.net/Applications

EXAMPLE:

Robert Smith pays the following bills from his bank account each month:

Electricity	£46·75
Gas	£21·25
Telephone	£42·50
Council tax	£169·00
Mortgage	£221·63

He estimates that he spends about £45 per week on food and he also puts aside £100 per week for entertainment and other expenses.

He works in a supermarket and his take-home pay was £1229·17 in February. Will he have a surplus or a deficit at the end of this month?

SOLUTION:

Expenditure from bank account
= £46·75 + £21·25 + £42·50 + £169·00 + £221·63 = £501·13

Extra expenditure = (£45 + £100) × 4 = £580

Total expenditure for February = £501·13 + £580 = £1081·13

He has a surplus because £1229·17 > £1081·13.

PLANNING AN EVENT

It is important that an accurate budget is prepared when planning events such as fund-raising concerts and events in aid of charities.

EXAMPLE:

Ms Chisholm is organising a school concert to raise money for charity.

The following costs will be incurred: advertising, £20; cost of printing tickets, £12·50; cakes and refreshments, £40; books of raffle tickets, £5·50; and raffle prizes, £30.

On the night of the concert, 175 tickets were sold at £6·50 each, 80 books of raffle tickets were sold at £1·25 each, the income from refreshments was £320 and there were donations amounting to £143.

How much money was raised for charity?

SOLUTION:

Expenditure = £20 + £12·50 + £40 + £5·50 + £30 = £108

Income = (175 × £6·50) + (80 × £1·25) + £320 + £143

= £1137·50 + £100 + £320 + £143

= £1700·50

Surplus = £1700·50 – £108 = £1592·50

The school concert raised £1592·50 for charity.

 THINGS TO DO AND THINK ABOUT

Do not use a calculator for question 1.

1. Marjorie Lee has a net salary of £1056·50 each month. She has estimated that each month she spends £210 on her mortgage, £138 on council tax, £113 on utility bills, £85 on her car and £275 on food and other expenses.

   a. Will Marjorie have a surplus or a deficit? **Justify** your answer with your working.

   b. Her car needs a repair that will cost £175 + value added tax (VAT). If the rate of VAT is 20%, what effect will this have on Marjorie's budget?

2. John Morrison has drawn a pie chart to show his monthly expenditure.

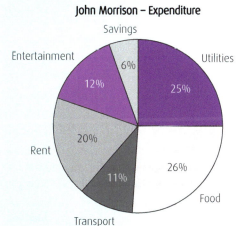

John Morrison – Expenditure

John's monthly take-home pay is £1200.
He is saving to buy a flat-screen television costing £500.
If he puts all his savings towards the television, how many months will it be before he can afford to buy it?

 ONLINE TEST

Test yourself on planning a budget at www.brightredbooks.net/Applications

EARNING MONEY 1

When you are in employment, you earn money. Your employment could be part-time or full-time. In full-time employment you are likely to be given an annual (yearly) salary and to be paid monthly. People can earn money in different ways. Some people have a basic salary that can be topped up. For example, a bus driver may be given an hourly rate of pay, which can be increased by working **overtime**. Other people may earn money through **piecework**, **commission** or **bonuses**. This section looks at all these different ways of earning money.

OVERTIME

The traditional working day is usually considered to be 9 am to 5 pm, five days a week from Monday to Friday. If employers want their employees to work at other times – for example, evenings and weekends – then this is called overtime and the basic hourly rate of pay is increased. The most common rates of overtime are double time (2 × basic wage) and time and a half (1·5 × basic wage). In this section, we consider three examples of overtime: a straightforward one; one involving working back; and one linked to the previous section on budgeting.

DON'T FORGET

To work out overtime at time and a half, multiply the basic rate by 1·5; if you are not allowed to use a calculator, half the basic rate and add it on – for example, at a basic rate of £7·80, time and a half is £7·80 + (£7·80 ÷ 2) = £7·80 + £3·90 = £11·70.

EXAMPLE:

Anna has a part-time job as a shop assistant. Her basic rate of pay is £7·80 per hour. She is paid time and a half for working on Sundays and double time for working on Bank Holidays.

If she works four hours on Thursday, three hours on Friday, eight hours on Saturday, two hours on Sunday and seven hours on Bank Holiday Monday, calculate her gross pay.

SOLUTION:

Pay at basic rate = (4 + 3 + 8) × £7·80 = £117

Overtime pay = 2 × (1·5 × £7·80) + 7 × (2 × £7·80) = £23·40 + £109·20 = £132·60

Gross pay = £117 + £132·60 = £249·60.

EXAMPLE:

Imran works a basic week of 38 hours. Any overtime worked is paid at time and a half.

One week he works for 45 hours and is paid £494·70.

How much is he paid for each hour of overtime he works?

SOLUTION:

Number of hours paid at basic rate = 38 + (45 − 38) × 1·5 = 38 + (7 × 1·5) = 48·5

Basic hourly pay = £494·70 ÷ 48·5 = £10·20

Overtime rate of pay = £10·20 × 1·5 = £15·30 per hour.

Sometimes you may wish to save up money from your wages to pay for an item you desire. This involves some budget planning as you will need to wait until you can afford the purchase.

contd

EXAMPLE:

Eve is saving up to buy the latest mobile phone. It costs £599. Each week she works four hours overtime on Friday night at time and a half and she works two and a half hours overtime on Saturday morning at double time. Her basic pay is £7·60 per hour. She knows that 30% of her gross pay goes on deductions.

If she saves her take-home overtime pay each week, how many weeks will it be before Eve can buy the mobile phone?

SOLUTION:

Gross overtime pay = 4 × (1·5 × £7·60) + 2·5 × (2 × £7·60) = £45·60 + £38 = £83·60

Deductions = 30% of £83·60 = 0·3 × £83·60 = £25·08

Take-home overtime pay = £83·60 − £25·08 = £58·52

Number of weeks = 599 ÷ 58·52 = 10·236

She will be able to buy the phone after 11 weeks.

PIECEWORK

People who make things are often paid for each item (or piece) that they make. In other words, the more they make, the more money they earn.

VIDEO LINK

Check out a video on overtime in the USA at www.brightredbooks.net/Applications

EXAMPLE:

Alexander works for a company that manufactures garden ornaments. He earns a basic salary of £75 per week, plus £6·50 for every ornament he makes.

Each week he saves 3/8 of his gross pay.

Calculate how much Alexander saves in a week in which he makes 18 ornaments.

SOLUTION:

Salary = £75 + (18 × £6·50) = £75 + £117 = £192

He saves 3/8 of £192 = £192 ÷ 8 × 3 = £72

THINGS TO DO AND THINK ABOUT

ONLINE TEST

Go to www.brightredbooks.net/Applications to test yourself on earning money.

Rashid earns £12·60 per hour for working a basic 40-hour week. For any extra hours he works, he is paid overtime at time and a half.

One week he works eight hours on Monday, seven and a half hours on Tuesday, eleven hours on Wednesday, nine and a half hours on Thursday and twelve and a quarter hours on Friday.

Calculate Rashid's gross pay for that week.

EARNING MONEY 2

COMMISSION

VIDEO LINK

Watch a video from the USA on how to calculate commission at www.brightredbooks.net/Applications

Employees who sell things in their job may earn commission as part of their salary. Commission is a percentage of the value of their sales. It rewards salespersons for doing their job well. There are three ways of earning commission. In straight commission, you only earn a percentage of what you sell (not good if sales are poor); in fixed salary + commission, you have a guaranteed salary (better if sales are poor); and in graduated commission, there are different rates of commission as sales increase (good if you sell a lot).

EXAMPLE:

Anna works in a call centre selling fitted bathrooms to customers.

Her pay is calculated as follows:

- for every customer agreeing to a home visit she is paid £12
- for every customer who buys a fitted bathroom she receives 1·5% commission.

During one month 43 of Anna's customers agreed to a home visit. They ordered £166 000 worth of bathroom fittings.

Calculate Anna's pay for that month.

SOLUTION:

Monthly pay = (43 × £12) + 1·5% of £166 000 = (43 × £12) + (0·015 × £166 000)

$$= £516 + £2490$$

$$= £3006$$

EXAMPLE:

Derek is paid a basic annual salary plus commission on his sales as shown in the table below.

Sales	Rate of commission
Less than £50 000	1·25% of all sales
£50 000 to £200 000	2·00% of all sales
More than £200 000	3·25% of all sales

His basic annual salary is £11 000.

What would Derek's sales need to be to achieve a total annual salary of £50 000?

SOLUTION:

His commission would have to be £50 000 − £11 000 = £39 000.

You must now check the different rates at the upper threshold to find the required rate:

1·25% of £50 000 = 0·0125 × £50 000 = £625

2·00% of £200 000 = 0·02 × £200 000 = £4000

His sales must be greater than £200 000 as £39 000 > £4000, therefore the required rate is 3·25%.

Hence 3·25% = £39 000 of his sales $\Rightarrow$ 1% = £39 000 ÷ 3·25 = £12 000.

Hence 100% of his sales = 100 × £12 000 = £1 200 000.

His sales would have to be £1 200 000.

DON'T FORGET

In commission questions where you have to calculate sales, work out 1% first (by division), then multiply by 100. It is a good idea to also check whether your answer is correct.

EARNING A BONUS

Some employees can top up their basic wage by earning a **bonus**. This may be given for meeting targets, for producing exceptional work, or for completing a job ahead of schedule.

EXAMPLE:

Joe Baxter is a professional footballer.

His basic weekly wage is £5600.

Each time his team wins a match, he earns a bonus of £400.

If his team wins a competition, he earns a bonus of 5% of his basic annual salary.

During a four-week period, Joe's team won three league matches and the cup final.

Calculate Joe's earnings over this four-week period.

SOLUTION:

Basic wage = 4 × £5600 = £22 400

Bonus for winning matches = 4 × £400 = £1600

Bonus for winning a trophy = 5% of (52 × £5600) = 0·05 × 52 × £5600 = £14 560

Earnings over four-week period = £22 400 + £1600 + £14 560 = £38 560

THINGS TO DO AND THINK ABOUT

1. Andrew is a car salesman. His monthly salary is £1450 plus 2·25% commission on all his sales. How much did he earn in a month when his sales were £152 000?

2. Anita sells cosmetics at parties in people's homes. The following table shows how her commission is calculated:

Sales	Rate of commission on sales
The first £500	20%
The next £250	30%
Any remaining sales	40%

   How much does Anita earn from a party if she sells £1650 worth of cosmetics?

3. Stephen is employed to make paperweights in a glass factory.
   He works a 40-hour week and is paid £7·20 per hour.
   His weekly target is to produce 150 paperweights.
   For each paperweight he produces above this target, he is paid a bonus of £2·40.
   One week Stephen makes 213 paperweights.
   Calculate Stephen's pay for that week.

ONLINE TEST

Go to www.brightredbooks.net/ Applications to test yourself on earning money.

INCOME TAX AND OTHER DEDUCTIONS

WHAT ARE DEDUCTIONS?

Each year the Chancellor of the Exchequer announces the government's budget for the nation for the following year. The government raises much of its spending through taxes such as income tax and National Insurance.

Chart 1: Government spending 2014–15

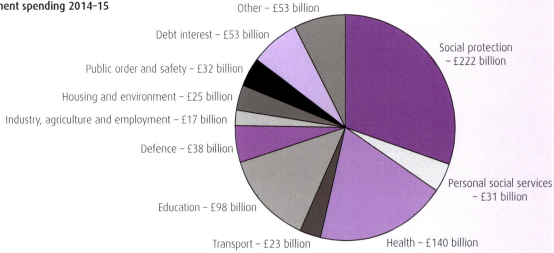

- Other – £53 billion
- Debt interest – £53 billion
- Public order and safety – £32 billion
- Housing and environment – £25 billion
- Industry, agriculture and employment – £17 billion
- Defence – £38 billion
- Education – £98 billion
- Transport – £23 billion
- Social protection – £222 billion
- Personal social services – £31 billion
- Health – £140 billion

Source: Office for Budget Responsibility. 2014–2015 estimates. Allocations to functions are based on HM Treasury analysis.

ONLINE

Go to www.brightredbooks.net/ Applications to find out more about the pie chart.

When you earn a salary, your **gross pay** is your salary before **deductions** are taken off. What remains is called your **net pay** (or take-home pay). Income tax and National Insurance are two deductions that are paid by everyone whose salary is above a certain amount. Other possible deductions may include superannuation (a pension scheme) and union fees. This section investigates income tax in some detail.

INCOME TAX

The rates of income tax can change from year to year. This section uses the income tax arrangements for the financial year 2015–16.

The standard personal allowance for 2015–16 is £10 600.

Taxable Income (£)	Rate (%)
£0 – £31 785	20%
£31 786 – £150 000	40%
Over £150 000	45%

EXAMPLE:

Hamish earns £48 750 per annum. He is entitled to the standard personal allowance. How much income tax will Hamish pay?

SOLUTION:

Taxable income = Income – Allowances = £48 750 – £10 600 = £38 150

Income tax paid at the 20% rate = 20% of £31 785 = 0·2 × £31 785 = £6357

Income tax paid at the 40% rate = 40% of (£38 150 – £31 785)
= 0·4 × £6365 = £2546

Hence total income tax paid = £6357 + £2546 = £8903

DON'T FORGET

Taxable income = income – allowances; you do not pay any tax on allowances.

Deductions such as income tax are taken off an employee's salary at source (before he or she receives them). If you are self-employed, you must complete you own tax return by a certain date each year (or get an accountant to do it for you). If you are late with your tax return, you will face penalties. Anyone who has overpaid or underpaid income tax can have their allowance adjusted the following year to make up the difference.

contd

EXAMPLE:

David Bell pays tax at the basic rate of 20%. He receives the standard personal allowance of £10 600. He owes tax from the previous year and has been informed that his allowance is to be adjusted so that the amount owed can be collected over the year, as shown in the example on the right.

If David owes £173, calculate his new allowance.

SOLUTION:

Amount of deduction = £173 × $\frac{100}{20}$ = £865

New allowance = £10 600 – £865 = £9735

Tax owed from previous years

Suppose you owe £256 and you pay tax at 20%, this will be collected by carrying out the following calculation:

£256 (amount owed) × $\frac{100}{20}$

(rate of tax)

= £1280 (amount of deduction)

So by deducting £1280 from your allowances, we collect the £256 you owe (£1280 at 20% = £256)

WORKING BACK

We can work back from the amount of income tax paid to find the annual salary.

EXAMPLE:

Marie paid £5316 in income tax during the previous year.

If her allowance is £10 600, calculate her gross salary for the year.

SOLUTION:

Check the different rates at the upper threshold to find the required rate:

20% of £31 785 = 0·2 × £31 785 = £6357

She must pay tax at the 20% rate as £5316 < £6375.

Marie's taxable income = £5316 × $\frac{100}{20}$ = £26 580

Marie's gross income = £26 580 + £10 600 = £37 180

OTHER DEDUCTIONS

The rates for National Insurance in 2015–16 are listed in the following table.

Income	Class 1 National Insurance Rate
Under £8060	0%
£8060 – £42 380	12%
Over £42 380	2%

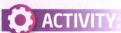 **ACTIVITY:**

In the case of Marie in the previous example, she would pay 12% of (£37 180 – £8060) in National Insurance during the year. Check this comes to £3494·40. Other deductions – for example, superannuation, are variable.

 THINGS TO DO AND THINK ABOUT

Using the given information on income tax rates and allowances, calculate the income tax paid in a year by:

1. Dorothy, who earns £15 750.
2. Ali, who earns £60 275.
3. Nigel, who earns £100 000.

 ONLINE TEST

Go to www.brightredbooks.net/Applications to test yourself on income tax and other deductions.

PAYSLIPS

WHAT DOES A PAYSLIP LOOK LIKE?

At the end of each month, employees are usually paid. Their net pay is paid into a bank account and they may receive a payslip. The payslip details personal information, all earnings during the month, any deductions and, finally, their net pay. Check the completed payslip in the following table to ensure you understand how net pay is calculated.

Name J. Singh	Employee No. 018	Tax code 1060L	Month May
Basic pay £1360·00	Overtime pay £212·50	Bonus –	Gross pay £1572·50
Nat. Insurance £108·10	Income tax £137·83	Pension £94·35	Deductions £340·28
			Net pay £1232·22

Basic pay + Overtime pay + Bonus = Gross pay

National Insurance + Income tax + Pension = Deductions

Net pay = Gross pay – Deductions

DON'T FORGET

Remember that
Net pay = Gross pay –
Deductions.

EXAMPLE:

James works in a leisure centre. Part of his duties is to prepare personal fitness plans for his clients. His April payslip is only partially completed:

Name J. Bond	Employee No. 007	Tax code 1060L	Month April
Basic pay £1284	Overtime pay –	Bonus	Gross pay
Nat. Insurance £102·73	Income tax £128·88	Pension	Deductions
			Net pay

James is paid a bonus of £3·75 for each client he works with.

During April, he worked on the personal fitness of 65 clients.

James pays 6% of his gross pay into his pension.

Calculate James's net pay for April.

ONLINE

Go to
www.brightredbooks.net/
Applications to check out
income tax and net pay using
an online tax calculator.

SOLUTION:

Bonus = 65 × £3·75 = £243·75

Gross pay = £1284 + £243·75 = £1527·75
Pension = 6% of £1527·75 = 0·06 × £1527·75 = £91·67
Deductions = £102·73 + £128·88 + £91·67 = £323·28
Net pay = £1527·75 – £323·28 = £1204·47.

contd

EXAMPLE:

A copy of Jordan Johnson's payslip is shown for the month of November:

Name J. Johnson	Employee No. 032	Tax code 1060L	Date 27/11/14
Basic pay £1435	Overtime pay £282·90	Bonus –	Gross pay £1717·90
Nat. Insurance £125·55	Income tax £166·91	Pension £103·07	Deductions £395·53
			Net pay £1322·37

Jordan worked 175 hours for his basic pay.

If overtime was paid at the time and a half rate, calculate how many hours of overtime he worked during November.

SOLUTION:

Basic hourly rate = £1435 ÷ 175 = £8·20
Overtime rate = £8·20 × 1·5 = £12·30
Number of hours of overtime worked = 282·90 ÷ 12·30 = 23.

EXAMPLE:

Dino sells software for an IT company. He earns a basic salary plus commission of 2·5% on all his sales.

His October salary slip is only partially completed:

Name D. Zoff	Employee No. 001	Tax code 1060L	Month October
Basic pay £1982	Overtime pay –	Commission	Gross pay
Nat. Insurance £181·84	Income tax £260·73	Pension £129·42	Deductions
			Net pay £1615·01

Calculate his sales for October.

SOLUTION:

Total deductions = £181·84 + £260·73 + £129·42 = £571·99
Gross pay = net pay + deductions = £1615·01 + £571·99 = £2187
Commission = £2187 – £1982 = £205
Hence 2·5% of his sales = £205 ⟹ 1% = £205 ÷ 2·5 = £82
Hence Dino's sales = 100 × £82 = £8200.

THINGS TO DO AND THINK ABOUT

Jean is a factory worker. Her payslip is only partially completed:

Name J. Campbell	Employee No. 177	Tax code 1060L	Month May
Basic pay £2300	Overtime pay £281·25	Bonus –	Gross pay A
Nat. Insurance C	Income tax £339·58	Pension £154·88	Deductions B
			Net pay £1857·64

What values should appear in boxes A, B and C in Jean's payslip?

 **DON'T FORGET**

Net pay is sometimes called take-home pay and is found by subtracting the deductions from the gross pay on a payslip. You should also realise that gross pay = net pay + deductions.

 ONLINE TEST

Go to www.brightredbooks.net/ Applications to test yourself on payslips.

FOREIGN EXCHANGE

Having worked for much of the year to earn money, it is common for people to enjoy holidays, sometimes abroad. When travelling abroad, they will have to convert British currency (pounds sterling) into foreign currency. The amount of foreign currency received depends on the exchange rate. This changes regularly and can be found online and in newspapers. You are expected to be able to convert between currencies, with at least three currencies involved in a multi-stage task.

EXAMPLE:

Pounds sterling (£)	Other currencies
1	$1·57 (US dollars)
1	$1·97 (Canadian dollars)
1	€1·41 (euros)

Alan goes on holiday to the USA. He buys a pair of sunglasses costing US$64.

a. How much is the pair of sunglasses in pounds sterling?

b. The same pair of sunglasses costs €62 in France and CAN$100 in Canada. In which of the three countries is the pair of sunglasses cheapest?

SOLUTION:

a. US$64 = 64 ÷ 1·57 pounds = £40·76433121 = £40·76 (to the nearest penny)

b. €62 = 62 ÷ 1·41 pounds = £43·97163121 = £43·97 (to the nearest penny)
CAN$100 = 100 ÷ 1·97 pounds = £50·76142132 = £50·76 (to the nearest penny)
The pair of sunglasses is cheapest in the USA as £40·76 is less than both £43·97 and £50·76.

EXAMPLE:

Tom changes £500 into euros. The exchange rate is £1 = 1·15 euros.

a. How many euros will he receive?
He spends 240 euros and changes the remaining euros to dollars at the rate of €1 = $1.12.

b. How many dollars will he receive? Give your answer to the nearest dollar.

SOLUTION:

a. £500 = 500 × 1·15 euros = €575

b. Number of euros remaining = 575 − 240 = 335
€335 = 335 × 1·12 dollars = $375·2 = $375 (to the nearest dollar)

DON'T FORGET

To convert British currency (£) to foreign currency, multiply the number of pounds by the exchange rate; to convert foreign currency to British currency, divide the amount of foreign currency by the exchange rate.

ONLINE

Go to www.brightredbooks.net/ Applications to find out about the Eurozone and the countries in it.

contd

EXAMPLE:

Theo is going on holiday to Cyprus. He changes £800 to euros at an exchange rate of £1 = €1·43.

a. How many euros will he receive?

b. While in Cyprus, he goes on a cruise to Israel. He knows that the exchange rate is £1 = 5·94 Israeli shekels. He wants to change 50 euros to Israeli shekels. How many shekels will he receive? Give your answer to the nearest shekel.

SOLUTION:

a. £800 = 800 × €1·43 = €1144

b. €50 = 50 ÷ 1·43 pounds = £34·96503497 = £34·97
£34·97 = 34·97 × 5·94 shekels = 208 shekels (to the nearest shekel).

CASE STUDY

EXAMPLE:

Peter is planning a summer holiday to Spain.

He is investigating different ways of changing pounds sterling to euros.

At his local travel agent, he finds that the exchange rate is £1 = €1·418. The travel agent charges commission at a rate of 1%.

At his bank, he finds that the exchange rate is £1 = €1·407 with no commission.

Online, he finds a website offering euros for sale. The online rate is £1 = €1·423 plus a fixed charge of £12 on all orders under £1500.

Peter wishes to change £1250 to euros.

What is the most economic method for Peter to buy his euros?

SOLUTION:

At the travel agent, commission = 1% of £1250 = £12·50, therefore number of euros = (1250 – 12·50) × 1·418 = €1755 (to the nearest euro).

At the bank, number of euros = 1250 × 1·407 = €1759 (to the nearest euro).

Online, number of euros = (1250 – 12) × 1·423 = €1762 (to the nearest euro).

It is most economic for Peter to buy his euros online as 1762 is greater than both 1755 and 1759.

DON'T FORGET

A case study is a longer example in which several pieces of information have to be considered and a conclusion reached. This example anticipates the section on 'Finding the best deal'.

ONLINE TEST

Go to www.brightredbooks.net/ Applications to test yourself on foreign exchange.

THINGS TO DO AND THINK ABOUT

Helen is going on a two-centre holiday to Hungary and Austria. She changes £1500 to Hungarian forints.

Pounds sterling (£)	Other currencies
1	436·88 HUF (Hungarian forints)
1	€1·41 (euros)

a. How many Hungarian forints will Helen receive?

b. Helen spends 450 000 HUF while in Hungary. If she changes all the remaining forints to euros for her trip to Austria, how many euros will she receive?

c. She spends €550 in Austria. How much money does she have left at the end of her trip? Give your answer in pounds sterling.

SPENDING MONEY

Spending money is an everyday part of life, whether it is a trip to the supermarket to buy fruit and vegetables or, at the other end of the scale, buying a house. It is important that you are familiar with terms such as discount, VAT, **hire purchase**, and profit and loss. Such ideas could feature as part of longer **case study** questions, so this section looks at different examples connected with spending money.

HIRE PURCHASE

EXAMPLE:

Patrick buys a washing machine using a hire purchase agreement.

The cash price of the washing machine is £480.

The total cost of buying the washing machine using hire purchase is 15% more than the cash price. Patrick pays a deposit of 20% of the cash price followed by 24 equal monthly repayments.

How much will Patrick pay each month?

SOLUTION:

Extra cost of hire purchase = 15% of £480 = 0·15 × £480 = £72

Total hire purchase cost = £480 + £72 = £552

Deposit = 20% of £480 = 0·2 × £480 = £96

Cost of repayments = £552 − £96 = £456

Cost of each payment = £456 ÷ 24 = £19.

VIDEO LINK

Watch a short clip on percentage profit and loss at www.brightredbooks.net/Applications

PROFIT AND LOSS

EXAMPLE:

Mr and Mrs King buy properties in poor condition, do them up, and then sell them on at a profit. They buy a house for £65 000, spend £10 000 on renovating the property, and later sell it on for £120 000. Express their profit as a percentage of their outlay.

SOLUTION:

Outlay = £65 000 + £10 000 = £75 000

Actual profit = £120 000 − £75 000 = £45 000

Percentage profit = $\frac{\text{actual profit}}{\text{outlay}} \times 100\% = \frac{45\,000}{75\,000} \times 100\% = 60\%$.

AT THE SUPERMARKET

Shoppers are often confronted by an array of special offers – for example, three for the price of two – at the supermarket. It can be difficult to decide which item offers the best value and may involve shopping around.

EXAMPLE:

A shop sells two bars of soap, each in the shape of a cuboid.

The larger bar costs £2·43. The smaller bar costs £1·28.

Which bar of soap gives better value for money?

Justify your answer.

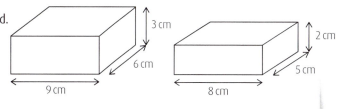

SOLUTION:

Calculate the volume of each bar and compare the costs.

Larger bar: $V = lbh = 9 \times 6 \times 3 = 162\,cm^3$ and cost = £2·43

Therefore $1\,cm^3$ of soap costs £2·43 ÷ 162 = £0·015

Smaller bar: $V = lbh = 8 \times 5 \times 2 = 80\,cm^3$ and cost = £1·28

Therefore $1\,cm^3$ of soap costs £1·28 ÷ 80 = £0·016

The larger bar is better value because £0·015 < £0·016.

DON'T FORGET

When you are asked to justify an answer, you should show all calculations, compare numerical values or state the difference between them, and reach a conclusion.

BUYING SHARES

It is possible to buy shares in companies. Most companies pay their shareholders a dividend each year. If the company does well, the share price may rise. You could then sell the shares and make a profit. However, the price of shares may also fall, so there is risk involved in buying shares.

EXAMPLE:

Jim buys 400 shares in a company for £5·25 each. When the price of each share rises to £7·20, he decides to sell all his shares. He is charged commission at 1·75% of the selling price when he sells the shares.

Calculate Jim's profit.

SOLUTION:

Cost price = £5·25 × 400 = £2100

Selling price = £7·20 × 400 = £2880

Commission = 1·75% of £2880 = 0·0175 × £2880 = £50·40

Jim's profit = £2880 − £2100 − £50·40 = £729·60.

ONLINE TEST

Go to
www.brightredbooks.net/
Applications to test yourself
on spending money.

THINGS TO DO AND THINK ABOUT

Chocolate bars are sold at a supermarket in multipacks of 6, 12 and 24. The prices of the multipacks are £2·88, £5·40 and £9·60, respectively.

a. Which pack offers the best value?

b. In a special offer, customers are offered three of the small packs for the price of two. How does this offer affect the best value? You must give reasons for your answers.

FINDING THE BEST DEAL

We have already considered this topic in the sections on foreign exchange (when we had to find the most economic method of changing pounds to euros) and spending money (when we had to find the best value bar of soap). Often this will involve using or calculating a unit rate – for example, the cost of one item or the cost of one litre.

VIDEO LINK

Check out a straightforward example at www.brightredbooks.net/Applications

COMPARING THREE PRICES

You are likely to be asked to determine the best deal by comparing at least three products, given three pieces of information on each. Questions of this type are case studies and involve having to read and understand numerous pieces of information. In order to do this successfully, you will require patience and organisation. Concentrate on one deal at a time because this will break the question down into more manageable parts.

EXAMPLE:

Arthur bought petrol last week at a cost of 118·5 pence per litre.

The petrol in his tank is now getting low and he decides to top it up by buying 36 litres of petrol before setting out on a business trip later in the day. He decides to drive from home to buy the petrol, then return home to have lunch before setting out on his trip.

There are three petrol stations he can choose from:

- petrol station 1 is 1·8 km from his home
- petrol station 2 is 7·2 km from his home and Arthur has a voucher for £2 off any purchase at this petrol station
- petrol station 3 is 9·6 km from his home and Arthur has a coupon for this petrol station that allows him 6 pence off per litre as long as he buys more than 30 litres.

The price of petrol has remained constant at 118·5 pence at all three petrol stations.

His car can travel 12 km on 1 litre of petrol.

Taking all factors into consideration, which petrol station should Arthur go to for the best value?

SOLUTION:

Petrol station 1:

Cost of petrol = 36 × 118·5p = 4266p = £42·66

Cost of petrol used on journey = (2 × 1·8) ÷ 12 × 118·5p = 35·55p = £0·36

Total cost = £42·66 + £0·36 = £43·02.

Petrol station 2:

Cost of petrol = 36 × 118·5p = 4266p = £42·66

Cost of petrol used on journey = (2 × 7·2) ÷ 12 × 118·5p = 142·2p = £1·42

Total cost (including voucher) = £42·66 + £1·42 – £2 = £42·08.

Petrol station 3:

Cost of petrol = 36 × (118·5 – 6)p = 4050p = £40·50

Cost of petrol used on journey = (2 × 9·6) ÷ 12 × 118·5p = 189·6p = £1·90

Total cost = £40·50 + £1·90 = £42·40.

Arthur should go to petrol station 2 as £42·08 is less than both £43·02 and £42·40.

DON'T FORGET

Questions on finding the best deal can be intimidating because there is a lot of information to read and it is easy to get confused. Break the task down into smaller chunks by focusing on only one of the deals at first. The working for the other deals is likely to be similar.

THINGS TO DO AND THINK ABOUT

ONLINE

Check out some price comparison websites for holiday travel, flights and accommodation.

1. Flora is looking for the best value of milk in her local supermarket.

   Which bottle gives best value?

| 2 litres £0·66 | 3 litres £0·92 | 4 litres £1·19 |

ONLINE TEST

Go to www.brightredbooks.net/ Applications to test yourself on finding the best deal.

   Justify your answer.

2. A shop sells three different kinds of paint.

COVERALL 5 litres £27·50   **GLOSSY** 4 litres £22·80   **EPAINT** 3 litres £18·30

   a. Explain why Coverall appears to give the best value for money.

   b. Further information is given on the back of each tin.

COVERALL 1 litre covers 14 m²   **GLOSSY** 1 litre covers 15 m²   **EPAINT** 1 litre covers 17 m²

   Using this additional information, decide which paint is the best value for money.

   Justify your answer.

INTEREST

Interest rates vary from time to time. Changes in interest rates affect most families in one way or another. If you are a saver and depend on the interest from bank accounts for your income, high interest rates are welcome; if you are a borrower paying back interest on a loan, such as a mortgage, low interest rates are welcome. We consider both cases in this section and deal with simple interest over a period of one year or less.

EARNING INTEREST

EXAMPLE:

Wendy deposits £2400 in her bank. The rate of interest is 2% per annum. How much interest will Wendy receive in 8 months?

SOLUTION:

Interest for a full year = 2% of £2400 = 0·02 × £2400 = £48

Interest for 8 months = £48 ÷ 12 × 8 = £32.

EXAMPLE:

Mr Mitchell is going to invest £15 000 in his bank for 1 year. He has checked the local branch of his bank to find their interest rates.

Name of account	Interest rates	Conditions
Current account	1·75% (taxed at 20%)	Plus a 0·1% bonus**
ISA	1·6%	Tax-free
Fixed-rate bond	1·96% (taxed at 20%)	**

** The bonus and fixed rate will only apply if there are no withdrawals during the year.

a. Assuming Mr Mitchell makes no withdrawals during the year, calculate the interest on each account.

b. Which account would you recommend Mr Mitchell to choose?

SOLUTION:

a. **Current account:**

Interest = (1·75 + 0·1)% of £15 000 = 0·0185 × £15 000 = £277·50

Tax = 20% of £277·50 = 0·2 × £277·50 = £55·50

Net interest = £277·50 − £55·50 = £222.

ISA:

Interest = 1·6% of £15 000 = 0·016 × £15 000 = £240.

Fixed rate bond:

Interest = 1·96% of £15 000 = 0·0196 × £15 000 = £294

Tax = 20% of £294 = 0·2 × £294 = £58·80

Net interest = £294 − £58·80 = £235·20.

b. I would recommend the ISA as £240 is greater than both £222 and £235·20.

DON'T FORGET

To calculate interest successfully, you must be expert in using percentages with and without a calculator. If in doubt, check out the section on percentages on pp. 78–79.

INTEREST ON LOANS

VIDEO LINK

Go to www.brightredbooks.net/Applications to watch a video about using the formula $I = Ptr$.

When you borrow money from a bank, building society or a loan company, you have to pay back what you borrowed plus interest on the loan. If you borrow a large amount – for example, a mortgage to pay for an expensive item such as a house – it can take years to pay everything back.

EXAMPLE:

Anika borrows £7200 to buy a car. She agrees to pay back the loan plus interest at the rate of 20% simple interest per annum.

a. Anika would like to pay off the loan over a period of 36 months. Calculate her monthly repayments.

b. She can only afford to pay £250 per month. Her brother offers to make up the difference each month. How much money will Anika's brother have to give her over the 36-month period?

SOLUTION:

a. Interest per annum = 20% of £7200 = 0·2 × £7200 = £1440

36 months = 3 years, so interest for 36 months = 3 × £1440 = £4320

Total amount to be repaid = £7200 + £4320 = £11520

Monthly repayments = £11520 ÷ 36 = £320.

b. Monthly difference = £320 – £250 = £70

Total amount given by brother = 36 × £70 = £2520.

THINGS TO DO AND THINK ABOUT

ONLINE TEST

Go to www.brightredbooks.net/Applications to test yourself on interest.

1. Josef deposits £12 500 in the bank. The rate of interest is 1·4% per annum. How much interest will Josef receive in nine months?

2. Mrs Lindsay wants to borrow £9000 to pay for a new bathroom. She can repay the loan over 12 months, 24 months or 36 months at a fixed simple interest rate of 10% per annum.

   a. Calculate the monthly repayments for each of the three loan options.

   b. Mrs Lindsay decides to pay off the loan over 12 months, but can only afford to pay £800 per month. Her mother offers to make up the difference each month. How much money will Mrs Lindsay's mother have to give her over the 12-month period?

BORROWING MONEY

REPAYMENTS

In the previous section, we introduced the idea of borrowing money and repaying the loan. In practice, the repayment calculations are more complicated. As a result, banks and finance companies prepare tables with details of the repayments for different loans and different periods of time.

DON'T FORGET

The cost of a loan is found by subtracting the amount loaned from the total repayments.

EXAMPLE:

The following table is used to calculate loan repayments at the Strathclyde Bank. It shows the monthly repayments on a loan of £1000.

APR (%)	12 months	24 months	36 months	48 months
10	£87·92	£46·14	£32·27	£25·36
12	£88·85	£47·07	£33·21	£26·33
14	£89·79	£48·01	£34·18	£27·33
16	£90·73	£48·96	£35·16	£28·34
18	£91·68	£49·92	£36·15	£29·37

James Milligan borrows £7500 over 24 months at an **annual percentage rate** (APR) of 16%. Use the table to calculate the cost of the loan.

SOLUTION:

Monthly repayment on £1000 = £48·96

Monthly repayment on £7500 = $\frac{7500}{1000}$ × £48·96 = £367·20

Total repayments over 24 months = 24 × £367·20 = £8812·80

Cost of loan = £8812·80 − £7500 = £1312·80.

EXAMPLE:

Donald borrowed £12000 from the Strathclyde Bank over a period of 48 months.

The loan cost him £2607·36.

Use the table to calculate the APR.

ONLINE

Go to www.brightredbooks.net/Applications to try out a loan repayment calculator.

SOLUTION:

Total repayments on £12000 loan = £12000 + £2607·36 = £14607·36

Total repayment on £1000 loan = £14607·36 ÷ 12 = £1217·28

Monthly repayment = £1217·28 ÷ 48 = £25·36

The APR is 10%.

PAYMENT PROTECTION

It is possible to buy payment protection insurance when taking out a loan. This can help you make your repayments if difficulties arise, such as illness or redundancy. There was a recent scandal over the mis-selling of such policies by some companies.

EXAMPLE:

The following table shows the monthly repayments charged by three companies for a loan of £20000 over a period of eight years, with and without payment protection.

Bank	With payment protection	Without payment protection
Strathclyde Bank	£359·75	£303·48
Tweeddale Bank	£367·47	£308·80
White Shark Loans	£392·71	£314·17

Rosemary takes out a £20000 loan from White Shark Loans, over eight years, with payment protection.

How much would she save by taking the loan without payment protection?

SOLUTION:

Monthly saving = £392·71 – £314·17 = £78·54

Total saving over eight years = (8 × 12) × £78·54 = £7539·84.

THINGS TO DO AND THINK ABOUT

1. The table below shows the monthly repayments to be made when £1000 is borrowed from Fair Deal Loans.

APR (%)	12 months	24 months	36 months
12	£94·82	£53·05	£39·18
14	£95·66	£53·88	£40·04
16	£96·59	£54·72	£40·89
18	£97·31	£55·54	£41·75

Arthur takes out a loan for £5500 over 36 months at an APR of 12%.

Find the cost of the loan.

2. Refer to the table given in the first example on the previous page.

Margaret needs a loan. She can afford to pay £250 per month and wants the biggest loan she can get over 24 months.

How much can she borrow from the Strathclyde Bank at 10% APR?

Give your answer to the nearest £100.

3. A building society has prepared the following table of monthly repayments for loans to a customer.

Amount of loan	Repayment over 1 year	Repayment over 2 years
£100	£9·03	£4·85
£200	£18·06	£9·70
£300	£27·09	£14·55
£500	£45·15	£24·25
£1000	£90·30	£48·50

The customer is told that 'The payments will be £108·36 per month if the loan is repaid over one year and £58·20 per month if the loan is repaid over two years.'

The customer wishes to repay the loan over two years. Calculate the cost of the loan.

ONLINE TEST

Go to www.brightredbooks.net/ Applications to test yourself on borrowing money.

CREDIT CARDS

UNDERSTANDING CREDIT CARDS

Credit cards can be used to purchase items in shops and stores, on the phone and online. No cash is required and the items are paid for later. Each card has a number, often 16 digits long, shows start and expiry dates, and has a security code on the back. The card should be signed on the back and a PIN number is given for use in shops and stores. Each credit card user is sent a monthly statement showing all recent transactions, the amount owed (the balance) and the minimum amount that must be paid back. If you only pay part of the balance, you will be charged interest on the remainder. This will be added to the following statement.

CREDIT CARD STATEMENTS

A simplified version of a **credit card statement** is shown below. Details of how to contact the credit card company, useful information and how to pay appear on the back of the statement.

VISA statement for account number 0123 4567 8901 2345
Name: Mr James Clitheroe

Date: 15 December 2016   Credit limit: £7500

Interest rate: 2·5% per month   Payment to arrive by 7 January 2017

28 November 2015	Previous balance	£240·00
4 December 2015	Payment – *Thank you*	80·00
	Amount due	160·00
	Interest	4·00
10 December 2015	Smooth Cosmetics	85·99
10 December 2015	Burger Palace	27·59
10 December 2015	Online Bookstore	13·48
	Balance due	291·06

Minimum payment: 3% of balance owed or £7·50 – whichever is greater

Note: Interest is charged each month on amount still owing after payment is deducted

VIDEO LINK

Go to www.brightredbooks. net/Applications for more information about reading a credit card statement.

In this statement, the credit limit of £7500 is for information only as the customer should not spend more than this amount in any given month. The previous balance (£240) is what was owed the month before. The amount paid (£80) has been deducted, so £160 is still owed. The interest of £4 is calculated as 2·5% of £160. The amount still owing, the interest and the three transactions on 10 December are added together to find that the balance to be paid this month is £291·06. There are details of the minimum payment the customer can make. Note that it is not a good idea to make only the minimum payment as you will still owe most of the balance, plus interest, the following month.

contd

EXAMPLE:

a. Mr Clitheroe makes the minimum payment. How much should he pay?

b. If Mr Clitheroe does not add any items to his credit card during the next month, calculate the 'balance owed' on the next statement.

SOLUTION:

a. 3% of £291·06 = 0·03 × £291·06 = £8·73

His minimum payment is £8·73 because £8·73 > £7·50.

b. On next statement:

Amount still owing = £291·06 − £8·73 = £282·33

Interest = 2·5% of £282·33 = 0·025 × £282·85 = £7·06

Balance owed = £282·33 + £7·06 = £289·39.

 DON'T FORGET

Make sure your answer seems sensible. Some students mistakenly include the credit limit in their calculations. Others make errors with the decimal point in the percentage calculation. They end up with answers of thousands of pounds. This would not be sensible after only three modest purchases, so check your working.

 THINGS TO DO AND THINK ABOUT

1. Christine has a four-digit PIN number for her credit card. She uses the digits from her date of birth, 6/2/97, but has forgotten the order. She can only remember that the first digit is 6.

   a. Write down all the four-digit pin numbers that Christine could try.

   b. If Christine has three consecutive incorrect attempts, she will be locked out of her account. She has two incorrect attempts. If she has another attempt, what is the probability that she will be locked out of her account?

2. Part of Alfredo's credit card statement is shown below. Calculate the values of A and B.

 ONLINE TEST

Go to www.brightredbooks.net/ Applications to test yourself on credit cards.

Alfredo Marconi

Credit limit = £2000

Balance from previous statement	£13·14
Interest	0·99
Smith's petrol station	49·66
Ed's mini-market	27·68
Trattoria Milano	52·05
Creative computers	69·18
Balance due	£A
Minimum repayment	£B

Minimum payment: 5% of balance due or £8 – whichever is greater

PROBABILITY

We cannot be certain what is going to happen when a particular event takes place – for example, tossing a coin. **Probability** gives a measure of how likely an event is to happen.

VIDEO LINK

Go to www.brightredbooks.net/ Applications and watch a video on simple probability.

PROBABILITY: AN OVERVIEW

Consider the event 'tossing a coin'. For every trial, there are two possible outcomes: heads or tails. We say that the probability of the coin landing heads is one in two (the probability of the coin landing tails is the same). We use the following definition for the probability of an event happening:

Probability of an event happening = $\frac{\text{number of favourable outcomes}}{\text{total number of outcomes}}$

- If an event is *impossible*, its probability is 0
- If an event is *certain*, its probability is 1
- Probability (event happening) + probability (event not happening) = 1
- Probabilities can *never* have either negative values or values greater than 1.

Probabilities are sometimes shown on a *probability line*.

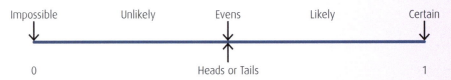

EXAMPLE:

The parks department in a city provides all its employees with suntan lotion for protection.

The department investigates how many of its employees are using the suntan lotion.

The results are shown in the table.

Job category	Using lotion	Not using lotion
Outdoor	20	4
Glasshouse	4	4
Clerical	1	3
Driver	7	5

What is the probability that an employee chosen at random does not use suntan lotion? Give your answer as a fraction in its simplest form.

SOLUTION:

Total number of employees = 20 + 4 + 4 + 4 + 1 + 3 + 7 + 5 = 48

Total number of employees not using lotion = 4 + 4 + 3 + 5 = 16

Probability = $\frac{\text{number of favourable outcomes}}{\text{total number of outcomes}} = \frac{16}{48} = \frac{1}{3}$.

contd

DON'T FORGET

Although probabilities are expressed in fractions in this section, they can appear in other forms, for example ½ could be written as 0·5 or 50% or as a one in two chance. Do not write 1:2.

EXAMPLE:

Two identical dice are rolled simultaneously.

Find the probability that the total score on adding both numbers will be eight.

SOLUTION:

Start by listing all possible outcomes when two dice are rolled.

1st die 2nd die	1	2	3	4	5	6
1	(1, 1)	(2, 1)	(3, 1)	(4, 1)	(5, 1)	(6, 1)
2	(1, 2)	(2, 2)	(3, 2)	(4, 2)	(5, 2)	**(6, 2)**
3	(1, 3)	(2, 3)	(3, 3)	(4, 3)	**(5, 3)**	(6, 3)
4	(1, 4)	(2, 4)	(3, 4)	**(4, 4)**	(5, 4)	(6, 4)
5	(1, 5)	(2, 5)	**(3, 5)**	(4, 5)	(5, 5)	(6, 5)
6	(1, 6)	**(2, 6)**	(3, 6)	(4, 6)	(5, 6)	(6, 6)

We can see that there are 36 possible outcomes when two dice are rolled. The favourable ones, a total of eight, are given in bold. There are five favourable outcomes, therefore the probability of both numbers adding to

$8 = \frac{5}{36}$.

EXPECTED FREQUENCY

If we toss a coin 100 times, the **expected frequency** of heads is 50. We can find the expected frequency for any outcome using the formula:

Expected frequency = number of trials × probability

EXAMPLE:

There are 1500 people in the audience at a concert.

The probability that a person is standing is $\frac{5}{12}$

How many people are standing?

SOLUTION:

Number standing = $1500 \times \frac{5}{12} = 625$.

 ONLINE TEST

Go to www.brightredbooks.net/ Applications to test yourself on probability.

THINGS TO DO AND THINK ABOUT

1. A lottery game uses black, white, red, blue and yellow balls. There are 20 balls of each colour numbered from 1 to 20. The balls are placed in a bag and one is drawn out.

   a. What is the probability that it is a 13?

   b. What is the probability it is a black 13?

PIE CHARTS

Pie charts provide a very attractive and clear way of illustrating data. In a pie chart, sectors of circles are used to show different pieces of information.

INTERPRETING A PIE CHART

EXAMPLE:

In a survey, a group of 240 girls was asked to choose their favourite sport. The results are illustrated in the pie chart.

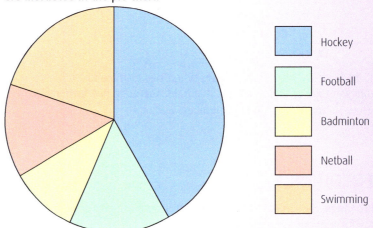

- Hockey
- Football
- Badminton
- Netball
- Swimming

How many of the girls chose hockey?

SOLUTION:

Measure the size of the angle for the hockey sector using a protractor. It is 150°.

Number of girls who chose hockey = $\frac{150°}{360°} \times 240 = 100$.

CONSTRUCTING A PIE CHART

EXAMPLE:

In the 2015 General Election, the percentage of votes cast for each political party in Scotland is given in the table.

Party	Percentage
Scottish National Party (SNP)	50·0
Labour (Lab)	24·3
Liberal Democrat (Lib Dem)	7·5
Conservative (Con)	14·9
Others	3·2

a. Construct a pie chart to illustrate this information. Show all of your working.

b. In the same election, the number of seats (out of a possible 59) won by each political party in Scotland is shown in the pie chart.

contd

Number of Scottish Seats 2015

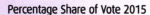

Some people claimed that the voting system was unfair. Comment on this claim.

SOLUTION:

a. Start by calculating, to the nearest degree, the angle for each sector:

Party	Working
SNP	50% of 360° = 360° ÷ 2 = 180°
Lab	24·3% of 360° = 0·243 × 360° = 87°
LD	7·5% of 360° = 0·075 × 360° = 27°
Con	14·9% of 360° = 0·149 × 360° = 54°
Others	3·2% of 360° = 0·032 × 360° = 12°

Now draw the pie chart moving clockwise from a line at 12 o'clock.
Label each sector and give the pie chart a title.

Percentage Share of Vote 2015

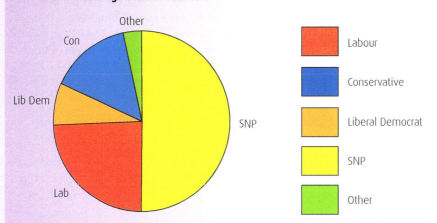

b. There is merit in this claim as the SNP won 56 out of 59 seats (95%) with only 50% of the votes cast. Similarly, Labour only won one out of 59 seats (2%), despite having 24·3% of the votes.

THINGS TO DO AND THINK ABOUT

1. A hotel is catering for 200 guests at a wedding.
   The pie chart, shown below, shows the choice of main courses.
   Calculate how many guests chose chicken.

2. 100 g of wholemeal bread contain 15 g of protein, 45 g of carbohydrates, 10 g of fibre, 5 g of fat and 25 g of other ingredients.
   Illustrate these data on a pie chart.

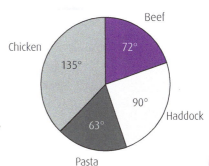

COMPARING DATA SETS

MEASURES OF CENTRAL TENDENCY

When a survey is carried out or an experiment takes place, sets of data are collected. Often, we illustrate the results in a graph or diagram. Another important part of statistics is to analyse the data sets or *distributions* and compare them with other data sets. This can be done by comparing some key measures from the data sets.

There are three measures of central tendency or *averages* which are used when comparing data sets. They are the **mean**, the **mode** and the **median**.

The mean = $\frac{\text{Total of all values}}{\text{Number of values}}$;

The mode = the most frequent value;

The median = the middle value in a set of ordered values.

> **EXAMPLE:**
>
> The results of 9 students in a test (out of 20) are listed below.
>
> 14   12   18   14   16   10   11   9   13
>
> Calculate (a) the mean;   (b) the mode;   (c) the median.

> **SOLUTION:**
>
> a. The mean = $\frac{\text{Total of all values}}{\text{Number of values}}$ = $\frac{14 + 12 + 18 + 14 + 16 + 10 + 11 + 9 + 13}{9}$ = $\frac{117}{9}$ = 13
>
> b. The mode = 14
>
> c. First order the data (lowest to highest) → 9 10 11 12 <u>13</u> 14 14 16 18
>
> The median = 13 (underlined as the middle value).
>
> Note that to find the median was easy, as it was obvious that 13 was the middle value. It is not always so easy, and in fact can become quite difficult when there is a large data set and when there is an even number of values. In the latter case, the median lies in between two values and is found by calculating the mean of these two values.

To help, we have a formula for calculating the position of the median in a list of ordered data.

In a set of ordered data with n values, the position of the median is $(n + 1) \div 2$.

Therefore in the last example, where $n = 9$, the position of the median could be found by using the formula leading to $(9 + 1) \div 2 = 5$. Hence the median was the 5th number in the ordered list.

> **EXAMPLE:**
>
> The percentage marks of 12 students in an exam are listed below.
>
> 43   57   62   56   70   38   43   59   80   63   35   91
>
> Calculate the median.

> **SOLUTION:**
>
> Order the data → 35   38   43   43   56   <u>57</u>   <u>59</u>   62   63   70   80   91
>
> As $n = 12$, position of median = $(12 + 1) \div 2 = 6\cdot5$. This means the median is between the 6th and 7th values (underlined). The median is therefore the mean of 57 and 59, which is calculated as $(57 + 59) \div 2 = 116 \div 2 = 58$. Hence the median = 58.

THE RANGE

While measures of central tendency tell us a lot about a data set, they do not tell us everything. Suppose a second group of 9 students had sat the same test (out of 20) as in the first example, and their marks were as follows.

| 12 | 13 | 13 | 13 | 13 | 13 | 13 | 13 | 14 |

If you calculate the mean, you will find that it is also 13, the same as the first group. If we were comparing the two groups, we could say that they had the same mean. However, we can see that the performances of the two groups were very different in other respects. To show this, we have a measure of spread for the data called the **range**.

The range = Highest value – Lowest value

For the two data sets with the same mean, the range is different each time (18 – 9 = 9 for the first set and 14 – 12 = 2 for the second set). The higher the range, the more spread out the data is, so we could conclude that the results in the second set were less spread out or were more consistent than those in the first set.

We use measures of central tendency and measures of spread to compare data sets.

EXAMPLE:

Some exam results are shown in a back-to-back stem and leaf diagram.

Boys   Girls

					5	2	**3**	9			
				6	5	1	**4**	8			
		9	8	6	2	**5**	0	4	6	9	
				7	5	3	**6**	6	6	7	7
					2	0	**7**	6	8	9	
						4	**8**	0	1		

n = 15   *n* = 15

| 3 | 9 | means 39 |
| 1 | 4 | means 41 |

a. Calculate   (i)   the median for the girls;
               (ii) the range for the girls.

b. Calculate   (i)   the median for the boys;
               (ii) the range for the boys.

c. Compare the performances of the girls and boys, and comment.

SOLUTION:

a. (i) As *n* = 15, position of median = (15 + 1) ÷ 2 = 8. The results are already ordered, so the median for the girls = 66.

  (ii) Range for girls = 81 – 39 = 42.

b. (i) As *n* = 15, position of median = (15 + 1) ÷ 2 = 8. The results are already ordered, so the median for the boys = 58.

  (ii) Range for boys = 84 – 32 = 52.

c. As the median for the girls is greater (66 > 58), on average, the girls seem to have performed better than the boys. The range for the boys is greater (52 > 42). This suggests that the results for the boys are more spread out.

DON'T FORGET

When comparing data sets, it is not enough to simply compare numbers and say that one is greater/smaller than another. You are expected to reach some kind of conclusion, for example who has done better and whose results are more spread out.

THINGS TO DO AND THINK ABOUT

Find the mean, mode, median and range for the following exam marks.

| 45 | 48 | 38 | 49 | 67 | 51 | 48 | 60 | 54 | 70 |

VIDEO LINK

Listen to the 'Mean, Median, Mode song' at www.brightredbooks.net/Applications

ONLINE TEST

Take the test 'Comparing Data Sets' online at www.brightredbooks.net/Applications

THE INTERQUARTILE RANGE AND BOXPLOTS

THE QUARTILES

The median divides a data set into two equal halves. The quartiles divide a data set into four equal quarters. To find the quartiles, first find the median of the data set. Then find the median of the first half, called the **lower quartile**. Then find the median of the second half, called the **upper quartile**.

We call the lower quartile Q_1, the median Q_2 and the upper quartile Q_3.

> **EXAMPLE:**
>
> Mrs Smith compares the price of a carton of milk in ten different shops.
>
> £1·09   £1·20   £1·05   £1·43   £1·00   £1·15   £1·25   £1·03   £1·19   £1·36
>
> Calculate (a) the median   (b) the lower quartile   (c) the upper quartile.

> **SOLUTION:**
>
> Order the data → 1·00 1·03 1·05 1·09 <u>1·15 1·19</u> 1·20 1·25 1·36 1·43
>
> As n = 10, position of median = (10 + 1) ÷ 2 = 5·5. This means the median is between the 5th and 6th values (underlined). The median is therefore the mean of 1·15 and 1·19, which is calculated as (1·15 + 1·19) ÷ 2 = 2·34 ÷ 2 = 1·17. Hence the median = £1·17.
>
> The lower quartile is the median of the lower half (1·00 1·03 1·05 1·09 1·15), that is £1·05, and the upper quartile is the median of the upper half (1·19 1·20 1·25 1·36 1·43), that is £1·25.
>
> Hence (a) Q_2 = £1·17   (b) Q_1 = £1·05   (c) Q_3 = £1·25.

THE INTERQUARTILE RANGE

In the previous section, we looked at the range as a measure of the spread of a data set. However, in certain cases, the range can be misleading. Suppose we were calculating the range for the data set 2, 25, 26, 28, 30, 32, 33, 85. The range (85 – 2 = 83) is affected by two values which 'lie outside' most of the other values in the data set. Values such as this are known as **outliers**.

A more useful measure of spread which focuses on the central numbers of the data set is the interquartile range. The interquartile range is the range of the middle half of a data set and is therefore unaffected by any outliers. The formula is given below.

$$\text{The interquartile range} = Q_3 - Q_1$$

In the example on the cost of milk, the interquartile range = £1·25 – £1·05 = £0·20.

Another measure of spread is the semi-interquartile range. It is half of the interquartile range and is also less affected by extreme values. The formula is given below.

$$\text{The semi-interquartile range} = \tfrac{1}{2}(Q_3 - Q_1)$$

> **EXAMPLE:**
>
> The ages of the members of a chess club are listed below.
>
> 18   23   56   26   46   23   36   70   32
>
> Calculate the interquartile range.

contd

SOLUTION:

Order the data → 18   23   23   26   32   36   46   56   70

As $n = 9$, position of median $= (9 + 1) \div 2 = 5$, hence the median (Q_2) = 32. The lower quartile is the median of 18, 23, 23, 26. Hence $Q_1 = (23 + 23) \div 2 = 46 \div 2 = 23$. This should be obvious without a calculation. The upper quartile is the median of 36, 46, 56, 70. Hence $Q_3 = (46 + 56) \div 2 = 102 \div 2 = 51$.

Hence the interquartile range $= Q_3 - Q_1 = 51 - 23 = 28$.

BOXPLOTS

A boxplot is a neat visual way of illustrating the key points of a data set. To draw a boxplot, we need a five-figure summary of the data set. This consists of the lowest value (L), the three quartiles and the highest value (H).

EXAMPLE:

Mr Ahmed times his car journey, in minutes, from his home to his workplace each day over a three-week period. The results are listed below.

20   25   18   26   21   35   30   26   25   29   31   18   35   30   27

a. Calculate (i) median   (ii) the lower quartile   (iii) the upper quartile.

b. Draw a boxplot to illustrate this data.

Mr Ahmed tries a new route to work over the next three weeks. He times each journey. The results for the new route are shown in the boxplot below.

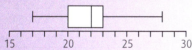

c. Compare the boxplots and make two appropriate comments about Mr Ahmed's journey times before and after he changed his route.

SOLUTION:

a. Order the data → 18 18 20 21 25 25 26 26 27 29 30 30 31 35 35

As $n = 15$, position of median $= (15 + 1) \div 2 = 8$, hence the median (Q_2) = 26. Check that Q_1 (the median of 18, 18, 20, 21, 25, 25, 26) = 21 and Q_3 (the median of 27, 29, 30, 30, 31, 35, 35) = 30. Hence the solutions are (i) 26   (ii) 21   (iii) 30.

b. The five-figure summary is L = 18, Q_1 = 21, Q_2 = 26, Q_3 = 30, H = 35.

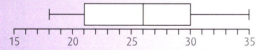

c. After he changed his route, on average, the journey time was less. We can tell this as the median fell from 26 minutes to 22 minutes, so the new route seems to be quicker. The boxplot shows that his journey times are less spread out after he uses the new route, so the journey times are more consistent.

DON'T FORGET

If you are asked to draw a boxplot, use a clear horizontal scale and make sure that you use a ruler. It is essential that boxplots are neatly drawn. If you are asked to compare two boxplots, then you should concentrate on two things, the median and the spread, and make appropriate comments in line with the advice from the previous section.

VIDEO LINK

Watch an excellent example of a boxplot taking shape with a comparison to follow at 'Box Plots': www.brightredbooks.net/Applications

ONLINE TEST

Take the test 'The Interquartile Range and Boxplots' online at www.brightredbooks.net/Applications

THINGS TO DO AND THINK ABOUT

For the data set: 24   16   17   25   20   34   31   13   28   27   26

a. Draw a boxplot.

b. Calculate the semi-interquartile range.

STANDARD DEVIATION

A MORE ACCURATE MEASURE OF SPREAD

We have met three measures which help us to see how a data set is spread out – the range, the interquartile range and the semi-interquartile range. We now consider a more accurate measure of spread called the **standard deviation**. Unlike other measures of spread, the standard deviation uses every member of the data set as part of the calculation. The symbol for standard deviation is s, and there are two formulae which can be used. They are shown below.

$$s = \sqrt{\frac{\Sigma(x - \bar{x})^2}{n-1}} = \sqrt{\frac{\Sigma x^2 - (\Sigma x)^2/n}{n-1}},\text{ where }n\text{ is the sample size.}$$

You will not have to memorise these formulae. However, take great care when copying them. We shall consider an example and use both formulae. Remember that $\bar{x}$ refers to the mean.

EXAMPLE:

A farmer delivers sacks of potatoes to supermarkets in his area. He checks the weight, in kilograms, of a sample of six sacks.

| 48 | 53 | 50 | 51 | 54 | 56 |

Calculate (a) the mean; (b) the standard deviation for this data set.

SOLUTION:

a. Mean = (48 + 53 + 50 + 51 + 54 + 56) ÷ 6 = 312 ÷ 6 = 52 kg

b. Method 1

$$s = \sqrt{\frac{\Sigma(x - \bar{x})^2}{n-1}} = \sqrt{\frac{42}{6-1}} = \sqrt{\frac{42}{5}} = \sqrt{8\cdot4} = 2\cdot9$$

(to 1 decimal place).

x	$x - \bar{x}$	$(x - \bar{x})^2$
48	48 − 52 = −4	16
53	53 − 52 = 1	1
50	50 − 52 = −2	4
51	51 − 52 = −1	1
54	54 − 52 = 2	4
56	56 − 52 = 4	16
		Total = 42

Method 2

$$s = \sqrt{\frac{\Sigma x^2 - (\Sigma x)^2/n}{n-1}} = \sqrt{\frac{16\,266 - 312^2 \div 6}{6-1}}$$
$$= \sqrt{\frac{42}{5}} = \sqrt{8\cdot4} = 2\cdot9$$

(to 1 decimal place).

x	x^2
48	2304
53	2809
50	2500
51	2601
54	2916
56	3136
Total = 312	Total = 16 266

Advice

You will probably prefer *one* of the above methods. Study the layout and the working carefully until you are confident you follow everything. Remember that the symbol Σ is called sigma and means 'the sum of'. Avoid a common mistake if using Method 1. When some students calculate $(x - \bar{x})^2$, they forget that when you square a number, the result is always positive. Many students have difficulty calculating $16\,266 - 312^2 \div 6$ in Method 2. You should do it all in one go on your calculator, write down the answer and *then* divide by the denominator (5 in this case). If you practise these skills, you should become an expert at this.

contd

EXAMPLE:

A sample of six boxes contains the following number of jelly beans per box.

28   27   29   30   32   28

a. For the above data, calculate:

   (i) the mean;   (ii) the standard deviation.

The manufacturers of the jelly beans claim that 'the mean number of jelly beans per box should be 30 (±2) and the standard deviation should be less than 2'.

b. Does the data in part (a) support the claim made by the manufacturers? Give reasons for your answer.

SOLUTION:

a. (i)   Mean = (28 + 27 + 29 + 30 + 32 + 28) ÷ 6 = 174 ÷ 6 = 29

   (ii) Method 1   Method 2

x	$x - \bar{x}$	$(x - \bar{x})^2$
28	28 – 29 = –1	1
27	27 – 29 = –2	4
29	29 – 29 = 0	0
30	30 – 29 = 1	1
32	32 – 29 = 3	9
28	28 – 29 = –1	1
		Total = 16

x	x^2
28	784
27	729
29	841
30	900
32	1024
28	784
Total = 174	Total = 5062

$$s = \sqrt{\frac{\Sigma(x - \bar{x})^2}{n - 1}} = \sqrt{\frac{16}{6 - 1}} = \sqrt{\frac{16}{5}} = \sqrt{3 \cdot 2} = 1 \cdot 8 \text{ (to 1 decimal place) or}$$

$$s = \sqrt{\frac{\Sigma x^2 - (\Sigma x)^2/n}{n - 1}} = \sqrt{\frac{5062 - 174^2 \div 6}{6 - 1}} = \sqrt{\frac{16}{5}} = \sqrt{3 \cdot 2} = 1 \cdot 8 \text{ (to 1 decimal place)}.$$

b. The manufacturers claim that the mean should be 30 (±2). This means between 28 and 32. The mean is 29, which is between 28 and 32. The standard deviation is 1·8, which is less than 2. So yes, the data does support the manufacturers' claim.

RELATED DATA SETS

EXAMPLE:

A data set has mean 25 and standard deviation 3. Find the mean and standard deviation if:

a. each member of the original data set is increased by 5;

b. each member of the original data set is doubled.

SOLUTION:

a. mean = 30; standard deviation = 3   b. mean = 50; standard deviation = 6.

NOTE: If we add the same number to each member of a data set, the mean will increase by that amount, but the standard deviation will be unchanged as the spread will remain the same. If we multiply each member of a data set by the same number, then both the mean and standard deviation will be multiplied by that number.

THINGS TO DO AND THINK ABOUT

Find the mean and standard deviation of the data set 18, 23, 19, 16, 20 and 24.

SCATTERGRAPHS

A **scattergraph** is a statistical diagram that can be used to compare two sets of data. It is drawn by plotting a set of points on a coordinate grid. We can use the graph to look for a **correlation** between the two sets of data. The correlation can be:

- positive (as one data set increases, the other also increases — for example, height and weight);

- negative (as one data set increases, the other decreases — for example, speed and time).

If the graph shows a cloud pattern, then there is no correlation — for example, height and salary. When there is a correlation, we can draw a **best-fitting line** on the scattergraph.

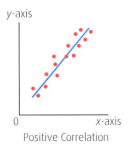

Positive Correlation

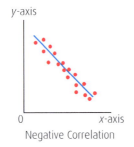

Negative Correlation

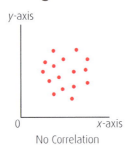

No Correlation

INTERPRETING A PIE CHART

EXAMPLE:

A teacher is investigating the marks of a group of students.

Their marks in their mathematics, physics and chemistry exams are shown in the table.

Student	A	B	C	D	E	F	G	H	I	J
Mathematics	15	46	66	42	70	38	60	49	77	85
Physics	20	45	60	32	57	43	68	56	65	89
Chemistry	29	36	43	28	60	35	57	48	70	65

The scattergraph shows the mathematics and physics marks of the group.

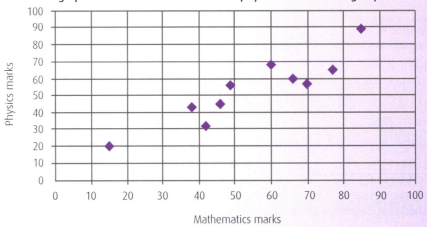

a. Draw a best-fitting line through the points on the graph.

b. Estimate the physics mark of student K, who scored 30 for mathematics.

c. Draw a scattergraph to show the mathematics and chemistry marks of the group and insert a best-fitting line on the graph.

d. Estimate the physics mark of student L, who scored 50 for chemistry.

contd

SOLUTION:

a.

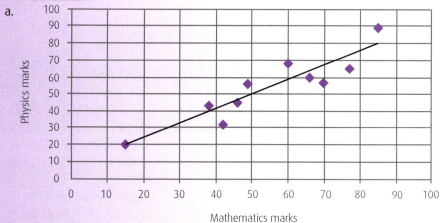

Physics marks / Mathematics marks

b. 32 (approximately)

c.

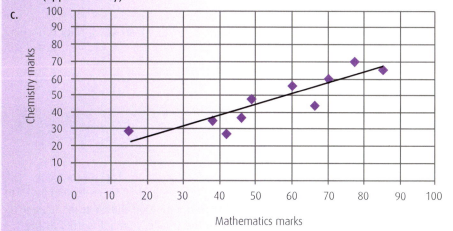

Chemistry marks / Mathematics marks

d. Use both graphs. Student L scored 50 for chemistry, leading to a mark of approximately 60 for mathematics. Then use the other graph to find an approximate physics mark of 59.

THINGS TO DO AND THINK ABOUT

After a study of taxi fares, the following table is produced.

Journey	A	B	C	D	E	F	G	H	I	J
Fare (£)	6	45	11	12	35	30	43	38	55	3
Distance (miles)	4	28	12	8	20	16	32	24	38	1

a. Illustrate these data on a scattergraph.

b. Draw a best-fitting line on the scattergraph.

c. Estimate the distance for a journey with a fare of £60.

GEOMETRY AND MEASURES

RELATED QUANTITIES 1

People come across **related quantities** in everyday life. Think of the cost per kilogram of potatoes in the supermarket, the number of kilometres per litre a driver can travel in a car, or the number of words per minute achieved by a typist. This and the following section consider related quantities and how to calculate one quantity based on two related quantities.

VIDEO LINK

Go to www.brightredbooks.net/ Applications to watch a video on direct proportion.

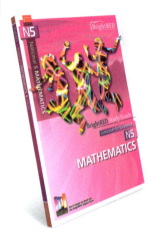

DIRECT PROPORTION

Two quantities are said to be in **direct proportion** if, as one quantity increases, the other increases at the same rate. For example, if a car can drive 40 miles per gallon of petrol, then it could drive 80 miles on two gallons.

EXAMPLE:

Mr McCallum buys 48 copies of a new mathematics textbook for £696. After the school roll increases, he orders a further 20 copies.

How much will the extra copies cost?

SOLUTION:

Number of copies	Cost
48	£696
20	$\frac{20}{48} \times £696 = £290$

Note: In this example, we could have calculated the cost of one copy of the book (£696 ÷ 48 = £14·50). This is called the unit cost and could then be multiplied by 20 to find the solution. However, it is convenient to use the method shown in the table in which £696 is reduced in the ratio $\frac{20}{48}$. Note that the smaller number is used as the numerator in the fraction because the cost of 20 copies will be less than the cost of 48 copies.

VIDEO LINK

Go to www.brightredbooks.net/ Applications to watch a video on indirect proportion.

INDIRECT PROPORTION

Two quantities are said to be in **indirect proportion** (or inverse proportion) if, as one quantity increases, the other decreases at the same rate. For example, if a journey takes four hours travelling at an average speed of 30 miles per hour, then it would take two hours travelling at an average speed of 60 miles per hour.

EXAMPLE:

It has been estimated that work on a building site can be completed in 48 days if 15 workers are employed. The building company will receive a bonus if the work is completed in 40 days. How many more workers should the company employ in order to earn the bonus?

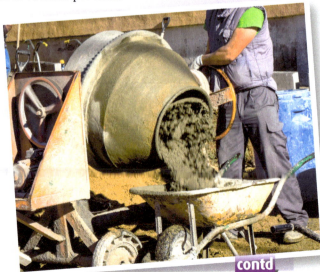

contd

SOLUTION:

Number of days	Number of workers
48	15
40	$\frac{48}{40} \times 15 = 18$

The company needs to employ three extra workers (18 – 15) to earn the bonus.

Note: In this example, we could have calculated the number of days required for one person working alone to compete the job (48 × 15 = 720). This is called the number of person-days required for the job and can then be divided by 40 to find the solution. However, it is convenient to use the method shown in the table in which 15 is increased in the ratio $\frac{48}{40}$. Note that the larger number goes in the numerator of the fraction because the number of people required will have to increase from 15 to complete the work more quickly. Note also that the product of the quantities (days and workers) is a constant (720).

DON'T FORGET

In proportion examples of the type shown, decide whether the related quantity has to be increased or decreased. For an increase, the larger number is the numerator of the fraction; for a decrease, the smaller number is the numerator of the fraction. Always check that your answer is sensible.

WHICH TYPE OF PROPORTION?

EXAMPLE:

Mike is saving up to buy a new laptop. If he saves £17·50 per week, it will take him 36 weeks to save up enough to buy the laptop. How much should he save each week if he wants to buy the laptop after 30 weeks?

SOLUTION:

He will need to save more than £17·50 per week to reduce the number of weeks, therefore this is an example of indirect proportion. To increase £17·50, the larger number is the numerator of the fraction.

Number of weeks	Amount saved
36	£17·50
30	$\frac{36}{30} \times £17·50 = £21$

 ## THINGS TO DO AND THINK ABOUT

1. It is estimated that a job will take 75 person-days to complete. Three people will start the job on Monday 7 March 2016.

   a. How many working days do they need to finish the job?

   b. If the three people work five days each week from Monday to Friday, on what day and date do they finish the job?

   c. At the end of the first week, how many person-days does the job still require?

   d. If a fourth person joins the workforce at the start of the second week, how many fewer days are then required to complete the job?

ONLINE TEST

Test yourself on related quantities at www.brightredbooks.net/Applications

RELATED QUANTITIES 2

REVERSING A PERCENTAGE CHANGE

Often a percentage is added to a sum of money – for example, VAT – or subtracted from a sum of money – for example, a discount. If we want to reverse the change that has taken place to find the original amount, we can do this using proportion.

EXAMPLE:

Alice buys a television. The total cost, including VAT at 20%, is £576.

What was the cost of the television before VAT was added?

SOLUTION:

The key to finding the solution is that £576 = 120% of the original price.

Percentage	Cost
120	£576
100	$\frac{100}{120} \times £576 = £480$

Hence the cost before VAT was £480.

If you are asked to solve a similar example without a calculator, you will have to be competent at using fractions.

EXAMPLE:

In a sale at a carpet store, a discount of $33\frac{1}{3}\%$ is offered on all marked prices. What was the marked price of a rug that was sold for £64?

SOLUTION:

The key to finding the solution is that
£64 = 100% − $33\frac{1}{3}\%$ = $66\frac{2}{3}\%$ of the marked price.

Hence the marked price was £96.

Percentage	Cost
$66\frac{2}{3}$	£64
100	$100/66\frac{2}{3} \times £64$ $= 300/200 \times £64$ $= 3/2 \times £64$ $= £64 \div 2 \times 3$ $= £96$

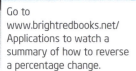

DON'T FORGET

In a reverse percentage problem, never calculate a percentage and subtract.

VIDEO LINK

Go to www.brightredbooks.net/ Applications to watch a summary of how to reverse a percentage change.

PROPORTION: USING A FORMULA

In the first example in the previous section, a textbook cost £14·50. We could use a formula to find the cost, C, of any number of copies, N. The formula would be $C = 14·5N$ in which the two quantities are related by a constant. We can always find a formula relating the quantities in proportion questions.

EXAMPLE:

The amount of tarmac, T tonnes, required to surface a road is directly proportional to the area, A square metres, of the road.

It takes 72 tonnes of tarmac to resurface a road with an area of 800 square metres.

a. Find a formula relating T and A.

b. How many tonnes of tarmac are required to resurface a road with an area of 2200 square metres?

contd

SOLUTION:

a. The quantities are related by the formula $T = kA$, where k is a constant.

Hence $72 = k \times 800 \Rightarrow k = 72 \div 800 = 0{\cdot}09$

The formula is $T = 0{\cdot}09\,A$.

b. $T = 0{\cdot}09\,A \Rightarrow T = 0{\cdot}09 \times 2200 = 198$

Hence 198 tonnes are required.

EXAMPLE:

A parachutist falls 125 metres during the first 5 seconds of her fall.

The distance, d metres, fallen by the parachutist is directly proportional to the square of the time, t seconds.

a. Find a formula connecting d and t.

b. Calculate the distance fallen by the parachutist in the first 7 seconds of her fall.

SOLUTION:

a. The quantities are related by the formula $d = kt^2$ where k is a constant.

Hence $125 = k \times 5^2$

$\Rightarrow 125 = k \times 25$

$\Rightarrow k = 125 \div 25$

$\Rightarrow k = 5$

Hence the formula is $d = 5t^2$.

b. $d = 5t^2$

$= 5 \times 7^2$

$= 5 \times 49$

$= 245$

The parachutist would fall 245 metres in 7 seconds.

 ## THINGS TO DO AND THINK ABOUT

 ONLINE TEST

Test yourself on related quantities at www.brightredbooks.net/Applications

1. Anne goes for a meal with some friends. The restaurant adds a 15% service charge to the bill. The total bill is £48·99.

   What was the price of the meal?

2. The weight of a steel bar, W grams (g), is directly proportional to its length, L centimetres (cm).

   A bar of length 16 cm weighs 200 g.

   a. Find a formula connecting W and L.

   b. Find the weight of a bar of length 26 cm.

SCALE DRAWINGS

REPRESENTATIVE FRACTION

Scale drawings form an important part of the work carried out by architects, builders and designers. To make a scale drawing, a suitable scale must be chosen. Maps show their scale by means of a **representative fraction**. The representative fraction is independent of units – for example, 1:1000. A representative fraction of 1:1000 means that one unit of measure on the map is equal to 1000 units of the same measure on the ground.

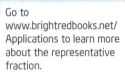

ONLINE

Go to
www.brightredbooks.net/
Applications to learn more
about the representative
fraction.

EXAMPLE:

Two towns are 5 cm apart on a map. The scale of the map is 1:250 000.

What is the actual distance between the two towns? Give your answer in kilometres.

SOLUTION:

Actual distance = 5 × 250 000 cm = 1 250 000 cm

As 1 km = 1000 m and 1 m = 100 cm, then 1 km = 100 000 cm

Hence actual distance = (1 250 000 ÷ 100 000) km = 12·5 km.

SCALE DRAWINGS

The **angle of elevation** of an object as seen by an observer is the angle between the horizontal and the line of sight from the observer to the object.

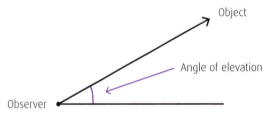

EXAMPLE:

An observer watches a hot air balloon from two different positions at X and Y.

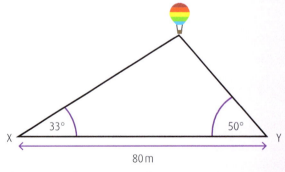

The distance between X and Y is 80 m.

The angle of elevation of the weather balloon is 33° from X and 50° from Y.

a. Make an accurate scale drawing to show the position of the balloon.

b. Use your scale drawing to find the actual height of the balloon.

contd

SOLUTION:

Choose a suitable scale: 1 cm represents 10 m (1:1000) is ideal.

Draw a base line of 8 cm to represent 80 m. Then use a protractor to measure angles of 33° and 50° and draw lines of sight for the angles of elevation until they meet. The dotted line shows the height of the balloon. Measure the height.

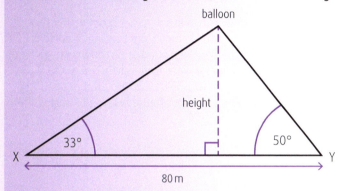

Height = 3·4 cm (approximately) ⇒ actual height of balloon = 3·4 × 10 = 34 m.

DON'T FORGET

Choosing a suitable scale is vital when making a scale drawing. If it is too large, it might not fit on the page; too small and the measurements will be less accurate. So think about the scale carefully before you start.

PLOTTING A COURSE

EXAMPLE:

A ship has to visit some oil rigs in the North Sea. The ship departs from its base. It sails 40 km due east, then 20 km due south and then 30 km due west to the oil rig Omega.

a. Make a scale drawing to show the ship's journey. Use a scale of 1 cm to represent 10 km.

b. The ship then returns directly from oil rig Omega to its base. Use your scale drawing to find the distance of this return journey.

SOLUTION:

Distances of 40 ÷ 10 = 4, 20 ÷ 10 = 2, 30 ÷ 10 = 3 should be used (in cm).

a.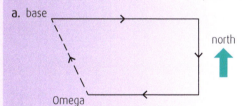

b. Use a ruler to measure the distance from Omega to the base (approximately 2·2 cm), then multiply by 10 for the solution in kilometres – that is, 10 × 2·2 = 22 km approximately.

 THINGS TO DO AND THINK ABOUT

A map has a scale of 1:1000. A loch is represented by a circle of radius 2 cm on the map. What area on the ground, in square metres, is represented by the area of the circle on the map? Use = 3·14 and do not use a calculator in this calculation.

 ONLINE TEST

Test yourself on scale drawings at www.brightredbooks.net/ Applications

NAVIGATION

THREE-FIGURE BEARINGS

Three-figure bearings are used to describe directions relative to north. The three-figure bearing of north is 000°. The three-figure bearings of other directions are given by angles measured clockwise from north. Hence the three-figure bearing of east is 090°, the three-figure bearing of south is 180° and the three-figure bearing of west is 270°.

EXAMPLE:

Monkland Farm is 4·5 km from the village of West Riding. Their relative positions are shown on the plan.

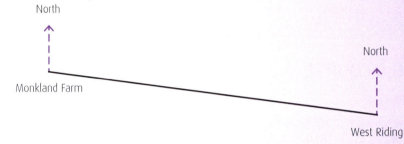

a. What is the scale of the plan?

b. What is the bearing of West Riding from Monkland Farm?

c. What is the bearing of Monkland Farm from West Riding?

SOLUTION:

a.

Distance on plan (cm)	Actual distance (km)
9	4·5
1	$\frac{1}{9} \times 4·5 = 0·5$

Scale is 1 cm = 0·5 km or 1:50 000

b. 096°

c. 276°.

contd

EXAMPLE:

A ship leaves port A and sails 43 km on a bearing of 075° to port B. It then changes course and sails 56 km on a bearing of 120° to port C.

a. Construct a scale drawing to show the ship's journey. Use a scale of 1 cm to represent 10 km.

b. The ship then sails back to port A. Use the scale drawing to find the distance and bearing of this journey.

SOLUTION:

a. From the starting point A, draw a north line. From A, measure the angle 75° and draw a line 4·3 cm long to represent 43 km. Mark point B and draw a north line. From B, measure the angle 120° and draw a line 5·6 cm long to represent 56 km. Mark point C.

b. The distance CA on the plan is 9·2 cm, hence the actual distance is approximately 92 km.

c. Use a protractor to find that the bearing of A from C is approximately 281°.

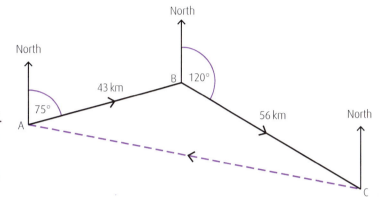

> **DON'T FORGET**
>
> Check the scale drawing carefully by measuring all lengths and angles.

THINGS TO DO AND THINK ABOUT

1. The bearing of ship A from ship B is 080°. The two ships are 40 km apart.

   a. Construct a scale drawing showing the positions of the two ships.

   b. Ship C is on a bearing of 135° from ship A and 218° from ship B. Plot its position on the diagram.

2. A ship leaves port P and sails 50 km on a bearing of 160° to port Q. It then changes course and sails 100 km on a bearing of 068° to port R.

   a. Construct a scale drawing to show the ship's journey. Use a scale of 1 cm to represent 10 km.

   b. The ship then sails back to port P. Use the scale drawing to find the distance and bearing of this journey.

> **ONLINE TEST**
>
> Test yourself on navigation at www.brightredbooks.net/Applications

CONTAINER PACKAGING

This section looks at the most efficient way of packing goods onto shelves or into containers. When packing goods for transportation — for example, furniture removal — it is important that the available space is used as efficiently as possible. Other factors may have to be considered because some goods must be kept upright for transportation.

EXAMPLE:

Eleanor is tidying her room. She is going to stack her DVDs on a shelf. She wants to stack as many DVDs on the shelf as possible.

She decides to stack the DVDs either vertically or horizontally, but not a mixture of both.

Each DVD is 19 cm high and 1·4 cm thick.

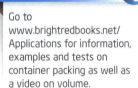
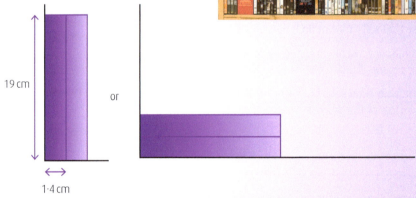

19 cm   or

1·4 cm

The shelf is 102 cm long and 22 cm high.

Which way should Eleanor choose to stack her DVDs?

Justify your answer.

SOLUTION:

Stacking vertically:

Number of DVDs across the shelf = 102 ÷ 1·4 = 72·857 = 72

Stacking horizontally:

Number of DVDs across the shelf = 102 ÷ 19 = 5·368 = 5

Number of DVDs one on top of the other = 22 ÷ 1·4 = 15·714 = 15

Number of DVDs = 5 × 15 = 75

Eleanor should stack the DVDs horizontally as 75 > 72.

contd

EXAMPLE:

Soup comes in cylindrical cans 12 cm high and with a diameter of 8 cm.

The cans are packed in boxes of length 48 cm, breadth 32 cm and height 24 cm.

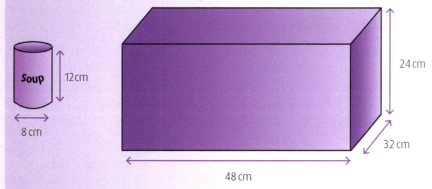

How many cans can be packed in a box?

SOLUTION:

Length: number of cans = 48 ÷ 8 = 6

Breadth: number of cans = 32 ÷ 8 = 4

Height: number of cans = 24 ÷ 12 = 2

Number of cans in box = 6 × 4 × 2 = 48.

DON'T FORGET

Many students make the error of trying to find the solution to this type of problem by calculating volumes. This is incorrect because there are empty spaces in the box. Always think of how the cans will fit into the space available. This will usually involve division.

THINGS TO DO AND THINK ABOUT

1. Stock cubes of side 2 cm are packed into a box in the shape of a cuboid with dimensions 8 cm × 6 cm × 4 cm.

   How many stock cubes can be packed into the box?

2. A company packs cylindrical cans of juice into boxes for delivery.

   Each can is 15 cm high with a diameter 10 cm.

   Each box is in the shape of a cuboid with length 80 cm, width 50 cm and height 60 cm.

   How many cans can be packed into each box if they are packed in an upright position?

3. Danny delivers goods to a supermarket in his van.
   The internal dimensions of the van are: length 3·5 m, width 2·2 m and height 1·8 m.

   He has to deliver boxes like the one shown below:

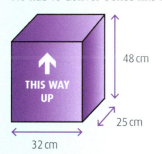

Find the maximum number of boxes Danny can pack into the van.

Justify your answer.

ONLINE TEST

Test yourself on container packaging at www.brightredbooks.net/Applications

PRECEDENCE TABLES

WHAT IS A PRECEDENCE TABLE?

If you look up 'precede' in a dictionary, you will find that it means 'to go before something in time, order or position' and precedence means 'priority'.

Suppose you plan to roast a chicken to eat with boiled potatoes. You would have to start by putting the chicken in the oven and roast it for, say, two hours. Later, you could cook the potatoes for about 20 minutes. This could be done while the chicken was cooking. So the total time for the preparation of the meal would be two hours.

A **precedence table** is a table that can be used to plan activities based on which activities must follow others and which activities can be carried out at the same time. Tasks are arranged in order from left to right and those tasks that can be carried out at the same time appear in the same column.

EXAMPLE:

Mario is about to start preparing a meal of spaghetti bolognese. The table shows the list of tasks and the time required for each.

Task	Detail	Preceding task	Time (minutes)
A	Prepare vegetables and grate parmesan	None	15
B	Soften vegetables in frying pan	A	9
C	Add mince to frying pan and brown	B	12
D	Add tomatoes and simmer meat sauce	C	45
E	Boil water in large pan	A	8
F	Cook spaghetti in pan	E	12
G	Drain spaghetti	F	2
H	Arrange spaghetti and sauce on plates	D, G	4
I	Add parmesan	H	1
J	Serve meal	I	3

a. Complete the diagram below by writing the tasks and times in the boxes.

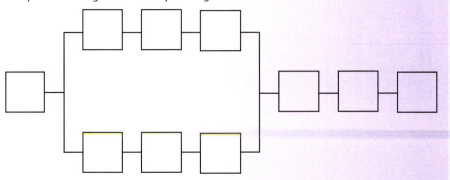

b. Mario wants to have the meal on the table at 5:15 pm. What is the latest time that he could start to prepare the meal?

contd

SOLUTION:

a.

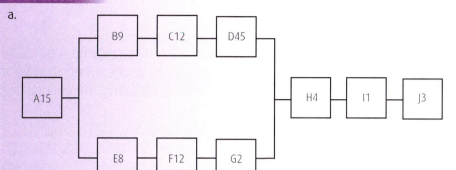

b. Add the times for each path, leading to ABCDHIJ = 89 and AEFGHIJ = 45. The longest path is called the **critical path** and this gives the length of time for the task to be completed. Thus Mario should start to prepare the meal 89 minutes before 5:15 pm at 3:46 pm.

THINGS TO DO AND THINK ABOUT

Samantha has decided to top up the amount of petrol in her car. The following list gives the tasks she will need to carry out:

A pay for petrol

B watch the cost

C drive to petrol station

D fill up tank

E drive home

F remove fuel cap

G switch off engine

H replace fuel cap.

Complete a precedence table by putting the activities in order. Identify which two activities could be carried out at the same time and put the second one in the right-hand column.

Order of activities	Activities that could be done at the same time

DON'T FORGET

When completing a precedence table, the start and finish should be obvious. Remember that tasks in the same column can be carried out simultaneously and look for any task that is preceded by two (or more) other tasks – for example, task H in this example.

ONLINE

Go to www.brightredbooks.net/Applications for information about precedence tables.

ONLINE TEST

Test yourself on precedence tables at www.brightredbooks.net/Applications

TIME MANAGEMENT

In the previous section on precedence tables, we saw an example of time management as Mario figured out how long it would take to prepare and serve the meal and to decide when to start the process. This section considers some more examples of managing time, including the importance of understanding timetables when planning a journey.

READING TIMETABLES

We all have to read timetables at some time when travelling on buses, trains, ferries and aeroplanes. Look at part of the train timetable from Glasgow Queen Street to Dundee.

Glasgow Queen Street *d*	0556	0741	0806	0841	0908
Larbert *d*	0616	0752	0820	0851	0920
Stirling *d*	0625	0809	0836	0908	0935
Dunblane *d*	0631	0815	0844	*	0941
Gleneagles *d*	0643	*	0856	*	0953
Perth *a*	0659	0841	0912	0939	1009
Perth *d*	0700	0842	0915	0940	1015
Invergowrie *d*	*	*	0939	*	1034
Dundee *a*	0722	0903	0944	1004	1043

Timetables are always given in 24-hour clock times. The vertical columns show the times of five morning trains from Glasgow Queen Street to Dundee. The letters *d* and *a* in the first column stand for departure and arrival – for example, the first train shown arrives in Perth at 0659 and departs one minute later at 0700. You will see a symbol (*) opposite certain stations. This means that the train does not stop there – for example, the first train shown does not stop at Invergowrie. Use this timetable in the next example.

EXAMPLE:

Sandi has to travel by train from Dunblane to Dundee for an interview at 10:40 am. Suggest which train Sandi should take from Dunblane. Justify your answer.

SOLUTION:

10:40 am = 1040 in 24-hour clock time, so Sandi must arrive in Dundee well before 1040. Arriving in Dundee at 1004 would probably be OK, but if we check the timetable for the train arriving in Dundee at 1004, we see that it does not stop at Dunblane. Therefore she needs to catch the earlier train leaving Dunblane at 0844.

TIME AND TEMPERATURE

When preparing a meal, you may have to pre-heat the oven before putting the dish in the oven. This has to be taken into account when working out meal preparation times.

EXAMPLE:

A recipe for beef stew gives the following instructions: 'Heat the oven to 230°C and cook the stew at this temperature for 25 minutes. Turn the switch to 130°C and leave the dish in the oven for a further 65 minutes.'

The temperature in the oven increases at a rate of 15°C per minute.

contd

a. Calculate how long it takes to make the beef stew if the oven starts at a temperature of 20°C.

b. If the stew has to be ready at 1900 hours, at what time should the oven be switched on?

SOLUTION:

a. Heating the oven takes (230 – 20) ÷ 15 = 14 minutes.

Total time = (14 + 25 + 65) minutes = 104 minutes.

b. Oven should be switched on at 104 minutes before 1900 – that is, at 1716 hours.

TIME ZONES

Different locations around the world have different time zones. This can cause difficulties for travellers and leads to the need to be careful when phoning back home so that you do not phone someone in the middle of the night.

EXAMPLE:

The opening ceremony of the London Olympics began at 2100 hours on Friday 27 July 2012. It was viewed by millions of people on televisions all over the world.

a. The time in Beijing, China is seven hours ahead of London time. What was the date and time of the start of the opening ceremony in Beijing?

b. The time in Rio de Janeiro, Brazil is four hours behind London time. What was the date and time of the start of the opening ceremony in Rio de Janeiro?

c. When it is 1635 hours in Beijing, what time is it in Rio de Janeiro?

d. A competitor in the London games decided to phone home to Beijing at 7:30 pm (London time). Is this a suitable time? Justify your answer.

e. The Chinese team flew home to Beijing on 13 August. The flight departed at 2025 and took 9 hours 45 minutes. What was the local time when the team arrived?

SOLUTION:

a. 0400 hours on Saturday 28 July 2012.

b. 1700 hours on Friday 27 July 2012.

c. 0535.

d. 7:30 pm in London is 2:30 am in Beijing. This is not a good time to phone as most people will be asleep.

e. 2025 + 9 hours 45 minutes + 7 hours = 1310 in Beijing.

 ## THINGS TO DO AND THINK ABOUT

Anita flies from Dubai to Glasgow for a meeting. The plane leaves Dubai at 1510. The flight takes 7 hours 45 minutes. Dubai time is three hours ahead of Glasgow time. When will Anita touch down in Glasgow?

 ONLINE TEST

Test yourself on time management at www.brightredbooks.net/Applications

TOLERANCE

Suppose we are told that the length of a piece of wood is 36 cm to the nearest centimetre. We can tell that the length has been rounded. Any measurement between 35·5 and 36·5 cm will round to 36 cm.

We say that 35·5 cm is the lower limit of a measure of 36 cm and 36·5 cm is the upper limit of a measure of 36 cm. This can be written in the form 36 ± 0·5 cm where the measurement has a **tolerance** of 0·5 cm on either side.

DON'T FORGET

There is an example involving tolerance in the section on standard deviation on p. 37. Remind yourself how to set out your answer to such examples.

EXAMPLE:

A company produces boxes of paperclips with an average content of 80 paperclips per box. As part of a quality control check, the contents of 30 boxes are recorded as:

77, 78, 80, 81, 78, 78, 79, 77, 76, 80, 78, 81, 85, 80, 79,

78, 82, 77, 80, 80, 81, 79, 78, 80, 78, 80, 83, 82, 80, 78.

Boxes with a content of 80 ± 2 paperclips are acceptable.

What percentage of the boxes is unacceptable?

SOLUTION:

80±2 means that boxes with contents between (80 − 2) = 78 and (80 + 2) = 82 inclusive are acceptable. Check the list carefully:

77, 78, 80, 81, 78, 78, 79, 77, 76, 80, 78, 81, 85, 80, 79,

78, 82, 77, 80, 80, 81, 79, 78, 80, 78, 80, 83, 82, 80, 78

There are six unacceptable boxes (shown in red).

Percentage of unacceptable boxes = $\frac{6}{30}$ × 100 = 20%.

EXAMPLE:

A delivery company is planning to stack three boxes, one on top of the other, in the back of a van. A space of 20 cm should remain above the boxes for access.

Each box should be placed upright and has a height of (45 ± 2) cm.

If the height of the interior of the back of the van is 1·6 m, can the boxes fit safely into the back of the van?

SOLUTION:

Upper limit of height of each box = (45 + 2) cm = 47 cm

Upper limit of height of all three boxes = 3 × 47 cm = 141 cm

Upper limit including gap = (141 + 20) cm = 161 cm

As the upper limit is 161 cm and the height of the van is 1·6 m (160 cm), it cannot be guaranteed that the boxes will fit in safely.

contd

You should be able to find the effect that tolerance has on measurements such as area and volume.

EXAMPLE:

The volume of a cuboid is 490 cm³.

The area of the base of the cuboid is 33 cm².

Both measurements are given correct to two significant figures.

Calculate the maximum height of the cuboid.

SOLUTION:

490 cm³ (to two significant figures) could range from 485 to 495 cm³ and therefore has a tolerance of (490 ± 5) cm³.

33 cm² (to two significant figures) could range from 32·5 to 33·5 cm² and therefore has a tolerance of (33 ± 0·5) cm².

The maximum height = $\frac{\text{upper limit of volume}}{\text{lower limit of area of base}}$ = $\frac{495}{32·5}$ = 15·23 cm (rounded to two decimal places).

DON'T FORGET

If you are asked to carry out a tolerance calculation involving division, remember that, for a maximum answer, divide the upper limit by the lower limit; for a minimum answer, divide the lower limit by the upper limit.

ONLINE

Go to www.brightredbooks.net/ Applications for information about tolerance.

THINGS TO DO AND THINK ABOUT

In a garden centre, plants are grown in the greenhouse. An employee keeps a record of the temperature of the greenhouse (in °C) at 12 noon every day for a week: 21, 22, 23, 19, 20, 17, 18.

a. Calculate the mean of the given temperatures.

b. For the best growth, the mean temperature should be 22 ± 2°C. Are the conditions in the greenhouse likely to result in the best growth? Justify your answer.

ONLINE TEST

Test yourself on tolerance at www.brightredbooks.net/ Applications

GRADIENT

The **gradient**, or slope, of a straight line is a measurement that tells us how steep the straight line is.

The gradient of a straight line can be calculated using the following formula:

Gradient = $\dfrac{\text{vertical height}}{\text{horizontal distance}}$

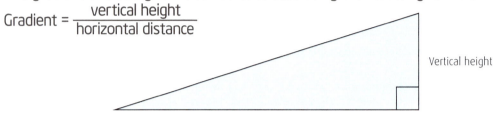

Vertical height

Horizontal distance

GRADIENT EXAMPLES

Ramps are used in many real-life situations – for example, to allow wheelchair access to buildings, to load large items onto trucks and vans, and even in skateboard parks. As a result of health and safety considerations, there are strict regulations about the gradient of such ramps.

EXAMPLE:

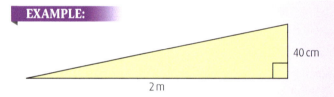

40 cm

2 m

Find the gradient of the ramp shown in the diagram. Express your answer as a fraction in its simplest form

SOLUTION:

The measurements must have the same units, so convert 2 m to 200 cm.

Gradient = $\dfrac{\text{vertical height}}{\text{horizontal distance}} = \dfrac{40}{200} = \dfrac{1}{5}$

EXAMPLE:

Safety regulations state that the maximum gradient on a railway network should be 0·027.

Measurements are taken on a particular slope on the network.

15 m

600 m

Does this slope meet the safety regulations? Give a reason for your answer.

SOLUTION:

Gradient = $\dfrac{\text{vertical height}}{\text{horizontal distance}} = \dfrac{15}{600} = 15 \div 600 = 0\cdot025$

Yes, the slope does meet the regulations because 0·025 < 0·027.

DON'T FORGET

To simplify a fraction, divide the numerator and denominator by their highest common factor. In this example, you can divide both 40 and 200 by 40, leading to the given solution.

DON'T FORGET

When you are asked to give a reason for your answer, you must answer by comparing the two numerical values or by stating the difference between them as well as reaching a conclusion. Answers such as 'Yes, because it is less' are unsatisfactory.

COORDINATES

When we consider gradient on a coordinate grid – for example, finding the gradient of the straight line joining two given points – some special rules apply.

- Parallel lines have the same gradient.

- Lines that slope upwards from left to right are said to have a positive gradient.

- Lines that slope downwards from left to right are said to have a negative gradient.

- Lines that are horizontal are said to have a zero gradient.

- Lines that are vertical are said to have an undefined gradient.

These results for gradient are summarised below.

Positive	Negative	Zero	Undefined
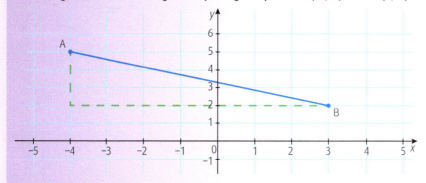			

When you are asked to find the gradient of the straight line joining two given points, you should plot the points, join them, form a right-angled triangle (if possible) and calculate the gradient using the formula given at the start of this section. You must also remember the given rules.

EXAMPLE:

Find the gradient of the straight line joining the points A (–4, 5) and B (3, 2).

SOLUTION:

Gradient of AB = $\frac{\text{vertical height}}{\text{horizontal distance}} = \frac{-3}{7} = -\frac{3}{7}$.

 ## THINGS TO DO AND THINK ABOUT

1. According to regulations, the gradient on a cycle track should be less than 0·05.

   A slope on a cycle track is shown below.

   12 m

   180 m

   Does this slope meet the safety regulations? Give a reason for your answer.

2. Find the gradient of the straight line joining the points A (3, 1) and B (9, 4).

COMPOSITE SHAPES 1

A **composite shape** is a shape made up of two or more shapes. The area of a composite shape can be found by adding (or subtracting) other areas. You should be able to use the following formulae:

Area of a triangle: $A = \frac{1}{2}bh$ Area of a circle: $A = \pi r^2$ Circumference of a circle: $C = \pi d$

VIDEO LINK

Watch a video on finding the area of composite shapes at www.brightredbooks.net/Applications

EXAMPLE:

Alexander's garden is in the shape of a symmetrical trapezium.

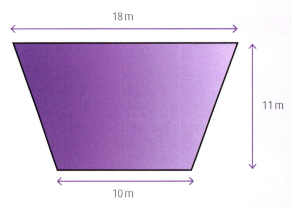

Alexander plans to sow grass seed in his garden. He knows that 40 g of grass seed are needed for each square metre of garden. Grass seed costs £8·75 per kg.

How much will it cost Alexander to sow grass seed in his garden?

SOLUTION:

Split the lawn into three parts as shown:

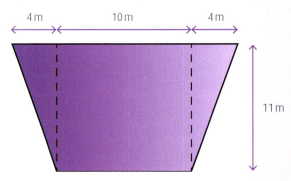

Area of rectangle: $A = lb = 10 \times 11 = 110$

Area of each triangle: $A = \frac{1}{2}bh = \frac{1}{2} \times 4 \times 11 = 22$

Area of lawn = $(110 + 22 + 22)\,\text{m}^2$ = 154 square metres

Number of kilograms of grass seed required for lawn = 154 × 40 ÷ 1000 = 6·16

Cost = 6·16 × £8·75 = £53·90.

EXAMPLE:

The badge for a school blazer is being designed. The badge is in the shape of a square and a semi-circle. The dimensions of the badge are shown below.

Calculate the area of the badge.

Give your answer correct to one decimal place.

6 cm

contd

SOLUTION:

To find the area of the badge, add the area of the square and the area of the semi-circle.

Area of square: $A = l^2 = 6^2 = 36$

Radius of semi-circle = $6 \div 2 = 3$

Area of semi-circle: $A = \frac{1}{2}\pi r^2 = 0{\cdot}5 \times \pi \times 3^2 = 14{\cdot}13716694$

Area of badge = $36 + 14{\cdot}13716694 = 50{\cdot}13716694$

Hence area of badge = $50{\cdot}1$ cm² (correct to one decimal place).

DON'T FORGET

It is better to use π rather than 3·14 in circle calculations as it gives a more accurate answer.

BEWARE: When carrying out calculations involving circles, it is common for students to mix up the formulae for the area and circumference. So be careful!

SECTORS OF CIRCLES

When two radii are drawn in a circle, two **sectors** are formed, called the major sector and the minor sector.

The area of a sector depends on the size of the angle subtended by the arc at the centre of the circle.
The relationship between the area of a sector and the area of the circle is:

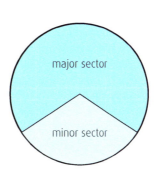

$$\frac{\text{area of sector}}{\text{area of circle}} = \frac{\text{angle at centre}}{360°}$$

This relationship can be used to find the area of a sector of a circle.

EXAMPLE:

The radius of the circle, centre O, is 11 cm.
Angle AOB = 103°.

Find the area of the minor sector OAB.

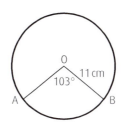

SOLUTION:

Area of sector = $\frac{103}{360} \times \pi r^2 = \frac{103}{360} \times \pi \times 11 \times 11 = 108{\cdot}76$

Hence area of sector = 110 cm² (correct to two significant figures).

ONLINE TEST

Test yourself on composite shapes at www.brightredbooks.net/ Applications

 ## THINGS TO DO AND THINK ABOUT

A standard athletics track is rectangular with semi-circular ends. Its dimensions are shown in the following diagram.

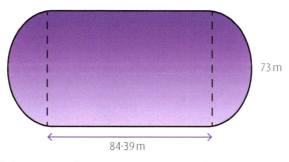

73 m

84·39 m

a. Calculate the area of the track (correct to the nearest square metre).

b. Steve plans to run 10 000 m around the perimeter of the track as part of his training. How many laps of the track will he have to run? Give your answer to the nearest lap.

COMPOSITE SHAPES 2

DECORATING: SOME BACKGROUND

Measurement and area are vital factors to be considered when decorating a room, whether the room is to be painted, wallpapered or carpeted. Many rooms are basically rectangular in shape, but most have little extra alcoves near doors or corners. Check the rooms in your own house as an example.

If you are painting a room, you need to calculate the area to be painted, taking into account doors, windows and skirting boards. Other factors to be considered are the type of paint – for example, gloss or emulsion – and how many coats you require.

If you are wallpapering a room, you will have to buy rolls of wallpaper. Suppose you buy rolls of wallpaper that measure $10\,m \times 50\,cm$. The wallpaper can be cut into strips depending on the height of the room. Therefore if a room is 3m high, you could cut off three strips from a 10m roll. These strips can then be pasted side by side onto the wall. Of course, that may not be possible if there is a pattern to be matched on the wallpaper. Carpets are also bought in rolls; however, fitting carpets is a fairly specialised task usually carried out by experts.

CASE STUDY

We now look at a longer case study on decorating a room. This involves calculating the areas to be painted or papered.

EXAMPLE:

Elizabeth is planning to decorate her room. The room is in the shape of a cuboid with the dimensions shown.

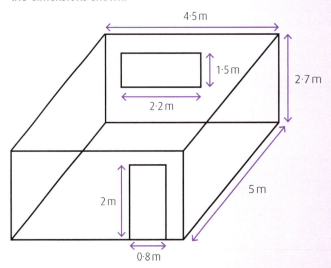

Information:

	Coverage	Cost
Emulsion	1 litre covers 16 m²	£20·50 for a 2·5 litre tin £8·80 for a 1 litre tin
Wallpaper	Each roll: Length 11 m Width 0·5 m	£29·95

contd

a. The wall with the window, the wall with the door and the ceiling are to be painted with one coat of emulsion paint. What is the minimum cost of the emulsion paint for the job?

b. The two plain walls are to be papered with wallpaper that requires no allowance for matching. Only full strips will be used.

   i. How many rolls of wallpaper are required?

   ii. What will be the cost of the wallpaper?

c. A tin of gloss paint costing £11·99 is bought to paint the woodwork. Calculate the total cost of decorating the room.

SOLUTION:

a. Area of front and back wall and ceiling = $2 \times (4\cdot5 \times 2\cdot7) + (5 \times 4\cdot5) = 46\cdot8\,m^2$

Area of window and door = $(2\cdot2 \times 1\cdot5) + (2 \times 0\cdot8) = 4\cdot9\,m^2$

Area to be painted = $46\cdot8\,m^2 - 4\cdot9\,m^2 = 41\cdot9\,m^2$

Elizabeth needs $41\cdot9 \div 16 = 2\cdot6\,l$ (to one decimal place).

The cheapest way to do this is to buy one 2·5 l tin and one 1 l tin of paint.

Cost of paint = £20·50 + £8·80 = £29·30.

b. Total length of walls to be papered = $2 \times 5\,m = 10\,m$

   i. Number of strips required = $10 \div 0\cdot5 = 20$

   Number of strips per roll = $11 \div 2\cdot7 = 4\cdot074 = 4$

   Number of rolls = $20 \div 4 = 5$.

   ii. Cost of wallpaper = $5 \times £29\cdot95 = £149\cdot75$.

c. Total cost of decorating room = £29·30 + £149·75 + £11·99 = £191·04.

DON'T FORGET

Bear in mind that, in real life, it is not simply enough to calculate the area to be papered because there will be some wastage. For example, when papering a room, allowance has to be made for matching, trimming and errors. You can find online calculators for paint, wallpaper and tiles that take these factors into account.

THINGS TO DO AND THINK ABOUT

A hotel is going to install decking on a patio around the garden. The patio, which includes a semi-circular area, is shown in the sketch.

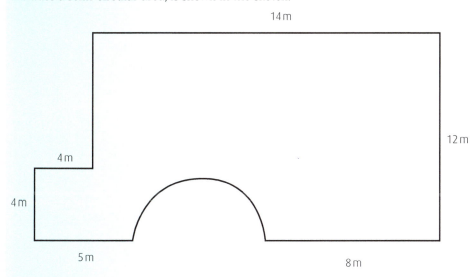

ONLINE

Check out an online calculator for decorating at www.brightredbooks.net/Applications

ONLINE TEST

Test yourself on composite shapes at www.brightredbooks.net/Applications

a. Calculate the area of the patio.

b. Decking cost £15·99 per m². As a result of the shape of the patio, the hotel buys 25% more decking than the required area. Calculate the cost for the hotel to install decking. Give your answer to the nearest £100.

VOLUME 1

RECAP

You should already know the formula for the volume of a prism:

Volume of a prism = area of base × height ($V = Ah$)

Three special types of prism are a cuboid, a cube and a cylinder:

Volume of a cuboid = length × breadth × height ($V = lbh$)

Volume of a cube = (length of side)³ ($V = l^3$)

Volume of a cylinder = area of base × height ($V = \pi r^2 h$)

You should also remember the basic units of volume, including the fact that 1 litre = 1000 cubic centimetres ($1l = 1000\,cm^3$).

VOLUMES OF STANDARD SOLIDS

We now look at the volumes of other standard solids, such as pyramids, cones and spheres.

The formula for the volume of a pyramid is $V = \frac{1}{3}Ah$, where A is the area of the base and h is the perpendicular height.

EXAMPLE:

Find the volume of the square pyramid shown in the diagram.

The pyramid has a perpendicular height of 15 cm and a square base of side 8 cm.

SOLUTION:

$V = \frac{1}{3}Ah = \frac{1}{3} \times 8 \times 8 \times 15 = 320$

Volume of pyramid = 320 cm³.

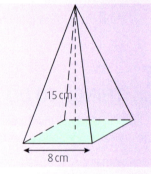
15 cm
8 cm

A special type of pyramid, with a circular base, is a cone. The formula for the volume of a cone is $V = \frac{1}{3}\pi r^2 h$.

EXAMPLE:

A cone has a height of 25 cm and a diameter of 30 cm.

Calculate the volume of the cone.

Give your answer correct to three significant figures.

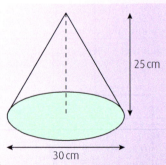
25 cm
30 cm

SOLUTION:

$V = \frac{1}{3}\pi r^2 h = \frac{1}{3} \times \pi \times 15^2 \times 25 = 5890{\cdot}486225$

Hence volume = 5890 cm³ (to three significant figures).

The formula for the volume of a sphere is $V = \frac{4}{3}\pi r^3$.

EXAMPLE:

A sphere has a diameter of 14 cm.
Calculate its volume.
Give your answer correct to three significant figures

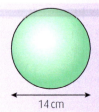
14 cm

contd

SOLUTION:

$V = \frac{4}{3}\pi r^3 = \frac{4}{3} \times \pi \times 7^3 = 1436\cdot755\,04$

Hence volume = 1440 cm³ (to three significant figures).
Note: You should also know how to use your calculator to find 7^3 without having to enter $7 \times 7 \times 7$.

WORKING BACK

On some occasions, you will know the volume of a standard solid such as a cylinder, cone or sphere. You may have to work back to find a missing dimension in the solid – for example, the height or the radius. Remember that the volume formulae for such solids involve multiplication, so, when you are working back, you will have to use the opposite operation, namely division. Study the following examples to see how this can be done.

EXAMPLE:

The volume of a cone is 1500 cm³.
Its height is 8 cm. Calculate its radius.

SOLUTION:

Write down the formula for the volume of a cone: $V = \frac{1}{3}\pi r^2 h$.

Now substitute the measurements given into the formula:

$1500 = \frac{1}{3} \times \pi \times r^2 \times 8$

Now find r^2 by division:

$r^2 = \dfrac{1500}{\frac{1}{3} \times \pi \times 8} = \dfrac{1500}{8\cdot377\,580\,41} = 179\cdot049.$

To find r, take the square root of r^2:

$r = \sqrt{179\cdot049} = 13\cdot38$

Hence the radius is 13 cm (correct to two significant figures).

Working back problems with a sphere are trickier. Suppose we know the volume of a sphere and want to find its radius. The formula for the volume of a sphere is $V = \frac{4}{3}\pi r^3$, therefore if we are working back, we will have to use the cube root to find the solution. Remember that the symbol for cube root is $\sqrt[3]{}$.

EXAMPLE:

A sphere has a volume of 5000 cm³. Find its radius.

SOLUTION:

$V = \frac{4}{3}\pi r^3 \Rightarrow 5000 = \frac{4}{3}\pi r^3 \Rightarrow r^3 = \dfrac{5000}{\frac{4}{3} \times \pi} = \dfrac{5000}{4\cdot188\,79} = 1193\cdot66$

$r^3 = 1193\cdot66 \Rightarrow r = \sqrt[3]{1193\cdot66} = 10\cdot607\,8$

Hence the radius = 10·6 cm (correct to three significant figures).

 DON'T FORGET

If you are working back to find the radius of a sphere, you need to calculate a cube root, so make sure you know how to do this on your calculator.

 ## THINGS TO DO AND THINK ABOUT

Give all answers correct to three significant figures.

1. Find the volume of a cylinder with a radius of 10 cm and a height of 35 cm.
2. Find the volume of a pyramid with a square base of side 9 cm and a height of 12 cm.
3. Find the volume of a cone with a diameter of 22 cm and a height of 19 cm.
4. Find the volume of a sphere with a diameter of 16 cm.
5. Find the height of a cylinder with a volume of 2500 cm³ and a radius of 10 cm.

 ONLINE TEST

Test yourself on volume at www.brightredbooks.net/Applications

VOLUME 2

We now look at some more complicated problems on volume. These will include composite shapes. A composite shape is made up of different parts — for example, a cylinder with a cone on top. In such cases, the volume needs to be calculated by addition.

In other situations, the volume of a shape may have to be calculated by subtracting different volumes. For example, the volume of a pipe may be found by subtracting the volume of one cylinder from another.

ADDING VOLUMES

EXAMPLE:

A barn shaped like a cuboid with a semi-cylindrical roof has the dimensions shown in the diagram.

Calculate the volume of the barn in cubic metres, giving your answer correct to 3 sig. figs.

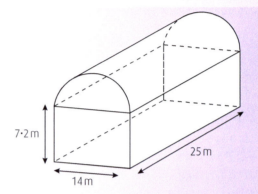

7·2 m

25 m

14 m

SOLUTION:

Volume of barn = volume of cuboid + volume of semi-cylinder

Volume of cuboid: $V = lbh = 25 \times 14 \times 7{\cdot}2 = 2520$

Volume of semi-cylinder: $V = \frac{1}{2}\pi r^2 h = \frac{1}{2} \times \pi \times 7^2 \times 25 = 1924{\cdot}2255$

Volume of barn = $2520 + 1924{\cdot}2255 = 4444{\cdot}2255$

Hence volume of barn = 4440 m³ (to three significant figures).

Note: This problem could also be solved using the formula for the volume of a prism: $V = Ah$. Calculate A, the area of the end of the barn (a rectangle + a semi-circle), then multiply by h, the length of the barn. This method is equally valid.

Check all your working carefully and remember not to round until the end.

SUBTRACTING VOLUMES

EXAMPLE:

a. An ice-cream tub is in the shape of part of a cone with the dimensions shown in the diagram.

Calculate the volume of the ice-cream tub.

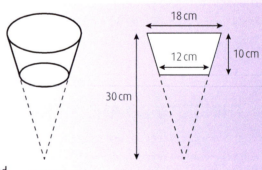

18 cm

12 cm

10 cm

30 cm

b. Another ice-cream tub is designed in the shape of a cylinder.

It has the **same** volume as the first tub. The diameter of this tub is 15 cm. Calculate the height of the tub.

15 cm

contd

SOLUTION:

a. Volume of ice-cream tub = volume of larger cone – volume of smaller cone

Volume of larger cone: $V = \frac{1}{3}\pi r^2 h = \frac{1}{3} \times \pi \times 9^2 \times 30 = 2544{\cdot}6900$

Volume of smaller cone: $V = \frac{1}{3}\pi r^2 h = \frac{1}{3} \times \pi \times 6^2 \times (30-10) = 753{\cdot}9822$

Volume of ice-cream tub = $2544{\cdot}6900 - 753{\cdot}9822 = 1790{\cdot}7078$

Hence volume of tub = $1791\,\text{cm}^3$.

b. Volume of cylindrical tub: $V = \pi r^2 h \Rightarrow 1791 = \pi \times 7{\cdot}5^2 \times h$

Hence $h = \dfrac{1791}{\pi \times 7{\cdot}5^2} = \dfrac{1791}{177} = 10{\cdot}1$

Hence the height of the tub is $10{\cdot}1\,\text{cm}$.

DON'T FORGET

Practise finding the volume of standard solids using the given formulae. Become familiar with the useful keys on your calculator, such as π, as well as the keys for roots and powers. Remember to halve the diameter if required to find the radius. Watch out for questions where the solution has to be rounded and do not round until the end.

VOLUME WITHOUT A CALCULATOR

EXAMPLE:

A sphere has a diameter of 6 cm.

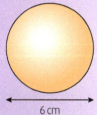

6 cm

Find its volume. Take $\pi = 3{\cdot}14$.

SOLUTION:

$V = \frac{4}{3}\pi r^3 = \frac{4}{3} \times 3{\cdot}14 \times 3^3 = \frac{4}{3} \times 3{\cdot}14 \times 3 \times 3 \times 3 = 113{\cdot}04$

Volume = $113{\cdot}04\,\text{cm}^3$.

Most students will find this calculation tricky without a calculator, but it is manageable if you approach it in the most efficient way. First, you must replace π with 3·14 and write 3^3 as $3 \times 3 \times 3$. Then remove the fraction (the cause of most of the difficulties) by cancelling it with the final 3 on the numerator, leaving $4 \times 3{\cdot}14 \times 3 \times 3$. Now multiply $4 \times 3{\cdot}14 = 12{\cdot}56$, then $12{\cdot}56 \times 3 = 37{\cdot}68$ and finally $37{\cdot}68 \times 3 = 113{\cdot}04$. To sum up, if you remove the fraction by cancelling, you can then do three fairly simple multiplications by 4, then 3, then 3 again. To show that you carried out all the calculations correctly, do not round your answer this time.

ONLINE

For more on calculating volumes, watch the tutorial at www.brightredbooks.net/Applications

 THINGS TO DO AND THINK ABOUT

A petrol tank is in the shape of a cylinder with hemispherical ends, as shown in the diagram. The total length of the tank is 1·8 m and the length of the cylinder is 1·3 m.

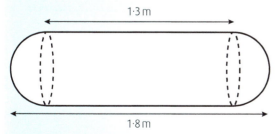

1·3 m

1·8 m

Calculate the volume of the tank in cubic metres.

Give your answer correct to two significant figures.

ONLINE TEST

Test yourself on this topic at www.brightredbooks.net/Applications

THE THEOREM OF PYTHAGORAS 1

PYTHAGORAS' THEOREM: A REMINDER

Pythagoras was a Greek mathematician who lived in the sixth century BC. He discovered an important fact about right-angled triangles. It is called **Pythagoras' theorem**. In words, the theorem states that:

In a right-angled triangle, the square on the hypotenuse is equal to the sum of the squares on the other two sides.

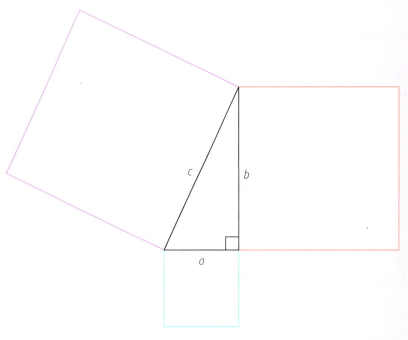

The **hypotenuse** is the longest side in a right-angled triangle and is opposite the right angle. A theorem is a true statement that can be expressed in words or by a formula. The formula for Pythagoras' theorem is:

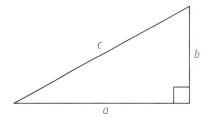

$$a^2 + b^2 = c^2$$

EXAMPLE:

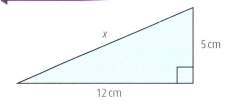

Find x.

SOLUTION:

Using Pythagoras' theorem, $x^2 = 12^2 + 5^2 = 144 + 25 = 169$

Hence $x = \sqrt{169} = 13\,\text{cm}$.

THE THEOREM OF PYTHAGORAS AND GRADIENT

EXAMPLE:

Regulations for ramps giving access to buildings state that: 'For ramps with a length between 5 and 10 m, the maximum gradient should be $\frac{1}{20}$.'

An existing ramp is 610 cm long and has a horizontal distance of 600 cm, as shown in the diagram.

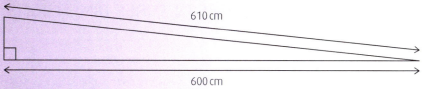

610 cm

600 cm

Does this ramp satisfy the regulations?

Justify your answer.

SOLUTION:

Use Pythagoras' theorem to find the vertical height, h, of the ramp:

$h^2 + 600^2 = 610^2$

$\Rightarrow h^2 = 610^2 - 600^2$

$\Rightarrow h^2 = 372\,100 - 360\,000$

$\Rightarrow h^2 = 12\,100$

$\Rightarrow h = \sqrt{12\,100} = 110$.

Gradient of ramp = $\frac{\text{vertical height}}{\text{horizontal distance}} = \frac{110}{600} = 0 \cdot 183$ (correct to three significant figures).

Maximum gradient = $\frac{1}{20} = 0 \cdot 05$.

The ramp does not satisfy the regulations as $0 \cdot 183 > 0 \cdot 05$.

TWO-STAGE CALCULATIONS

EXAMPLE:

A sketch of a component for a machine is shown in the diagram.

Calculate the length of side AC in the component.

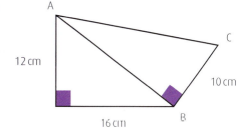

A

C

12 cm

10 cm

16 cm

B

SOLUTION:

$AB^2 = 16^2 + 12^2 = 256 + 144 = 400$

$\Rightarrow AB = \sqrt{400} - 20$

$AC^2 = 10^2 + 20^2 = 100 + 400 = 500$

$\Rightarrow AB = \sqrt{500} = 22 \cdot 4$ cm (correct to one decimal place).

THINGS TO DO AND THINK ABOUT

Aisha goes online to buy a tray to put on her coffee table.
She spots a rectangular tray with length 40 cm and width 25 cm.
Aisha's coffee table is circular with a diameter of 50 cm.

Will the tray fit on the coffee table? Justify your answer.

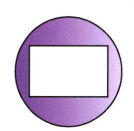

ONLINE TEST

Test yourself on the theorem of Pythagoras at www.brightredbooks.net/ Applications

THE THEOREM OF PYTHAGORAS 2

The theorem of Pythagoras can be used to solve a variety of problems involving three dimensions and circles.

THE THEOREM OF PYTHAGORAS IN THREE DIMENSIONS

EXAMPLE:

AB represents a flagpole at the corner of a field.

The flagpole is 8 m high. BCDE represents the field, which is rectangular.

BC is 9 m long and DC is 24 m long.

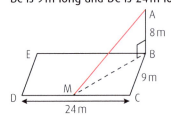

A rope stretches from A, the top of the flagpole, to M, the midpoint of DC.

Calculate the length of the rope.

SOLUTION:

We do two calculations involving the theorem of Pythagoras, first in triangle BCM and second in triangle ABM.

$BM^2 = 9^2 + 12^2 = 81 + 144 = 225 \Rightarrow BM = \sqrt{225} = 15$

$AM^2 = 15^2 + 8^2 = 225 + 64 = 289 \Rightarrow AM = \sqrt{289} = 17$

Hence the rope is 17 m long.

VIDEO LINK

Watch the first part of 'Lengths and Angles inside Cuboids' at www.brightredbooks.net/ Applications for a clear demonstration of how to find the length of the space diagonal in a cuboid using the theorem of Pythagoras.

SYMMETRY IN A CIRCLE

A chord in a circle is a straight line joining two points on the circumference. A line from the centre of a circle perpendicular to a chord bisects the chord. Similarly, the perpendicular bisector of a chord passes through the centre of a circle.

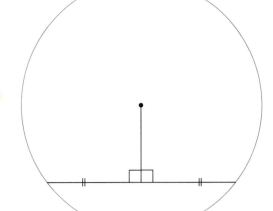

DON'T FORGET

The perpendicular bisector of a line is a straight line which crosses the midpoint of the line at right angles.

The presence of right angles in symmetrical diagrams in circles means that many problems can be solved using the theorem of Pythagoras.

contd

EXAMPLE:

The entrance to a tunnel is shaped like part of a circle.

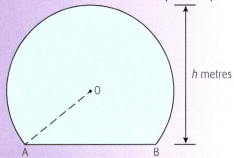

The radius of the circle, centre O, is 3·8 m. The length of the base of the entrance, represented by AB, is 6 m.

Calculate the height, h, of the entrance in metres.

SOLUTION:

This is the type of problem that can be solved using the theorem of Pythagoras. Join O to A (or B) and draw a perpendicular line from O to the midpoint of AB. This will form a right-angled triangle. The perpendicular line will bisect AB. Let the midpoint of AB be M. Therefore AM = 3 m.

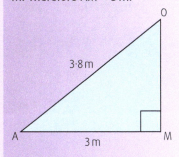

Hence $OM^2 = 3·8^2 - 3^2 = 14·44 - 9 = 5·44 \Rightarrow OM = \sqrt{5·44} = 2·33$.

To find h, the height of the tunnel, we must add the length from O to the top. As this length is the radius, $h = 3·8 + 2·33 = 6·13$.

Therefore the height of the tunnel is 6·1 m (to two significant figures).

 THINGS TO DO AND THINK ABOUT

1. A gift box is in the shape of a cuboid with dimensions as shown in the diagram.

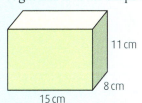

11 cm

8 cm

15 cm

Calculate the length of a space diagonal of the cuboid.

2. A tank is used to transport petrol.

The cross-section of the tank is a circle with diameter 6·4 m.

The surface of the petrol, represented by PQ in the diagram, is 5 m wide.

Calculate the depth, d, of petrol in the tank in metres.

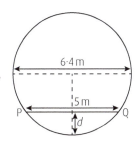

6·4 m

5 m

P   Q

d

NUMERACY

ROUNDING

APPROXIMATION

In real-life situations, numbers are often rounded to approximate values depending on the accuracy required. For example, we might measure our height to the nearest centimetre or the interest given by a bank may be rounded to the nearest penny. The population of the UK was 63 182 000 to the nearest thousand according to the 2011 census; it would be equally appropriate to say that the population was around 63 million.

The process of giving approximate answers involves rounding. This section looks at different ways of rounding, including significant figures.

EXAMPLE:

An example of rounding you should be familiar with relates to foreign currency.

Suppose you return from a holiday in Spain with 100 euros and want to change this money into pounds sterling. You find that the exchange rate is £1 = 1·174 euros. Convert the euros to pounds sterling, giving your answer to the nearest penny.

SOLUTION:

100 euros = 100 ÷ 1·174 pounds = £85·178 875 64 = £85·18 (to the nearest penny).

DECIMAL PLACES

In the previous example, we were effectively rounding the answer to two decimal places because we required two numbers after the decimal point in the solution. Using decimal places (the number of figures that appear after the decimal point) is a common way of rounding.

For example, 6·753 871 has six decimal places. Suppose you were asked to round it to three decimal places:

- 6·753 871 lies between 6·753 and 6·754

- look at the figure in the fourth decimal place (6·753 8) – that is, 8

- because this figure is more than 5, round up to 6·754.

Check that 6·753 871 rounded to two decimal places is 6·75.

DON'T FORGET

If the next figure is 5 or more, round up. If the next figure is 4 or less, do not round up.

EXAMPLE:

The radius of a circle is 15 cm. Calculate its circumference. Give your answer correct to one decimal place.

SOLUTION:

$C = \pi d = \pi \times 30 = 94\cdot247\,777\,961 = 94\cdot2$ cm (to one decimal place).

SIGNIFICANT FIGURES

Another method of rounding to an approximate number is to use significant figures (sig. figs for short). In a number, all figures are significant except the zeros that are used simply to indicate the position of the decimal point. However, zeros in between other significant figures are themselves significant.

Hence 589 346 has 6 sig. figs, 715·28 has 5 sig. figs and 4045 has 4 sig. figs. Note that 0·003 48 has only 3 sig. figs because the zeros at the start simply indicate the position of the decimal point, whereas 0·003 048 has 4 sig. figs because the zero between the 3 and 4 is significant. A measurement such as 25·0 cm has 3 sig. figs because the final zero tells you that it is a more accurate measurement than simply 25 cm.

Be careful with whole numbers. A crowd at a football match of 25 000 to the nearest thousand only has 2 sig. figs. However, when you say that there are 90° in a right angle, there are 2 sig. figs because the number is exactly 90. In other words, when there are trailing zeros in a whole number, it depends on whether the number has been rounded or is exact.

The rules for rounding remain the same – that is, round up if 5 or more, do not round up if 4 or less.

> **EXAMPLE:**
>
> Round
>
> a. 33 528·746 to 2 sig. figs.
>
> b. 0·002 797 to 3 sig. figs.
>
> **SOLUTION:**
>
> a. 33 528·746 (which has 8 sig. figs) lies between 33 000 and 34 000. As the third significant figure is 5, we round up to 34 000. Note: do not write down 34 000·0, which would have 6 sig. figs.
>
> b. 0·002 797 (which has 4 sig. figs) lies between 0·002 79 and 0·002 80. As the fourth significant figure is 7, we round up to 0·002 80.

> **EXAMPLE:**
>
> A cylinder has a height of 25 cm and a radius of 6 cm.
>
> Calculate its volume. Give your answer correct to 2 sig. figs.
>
> Volume of a cylinder = $\pi r^2 h$.
>
> **SOLUTION:**
>
> $V = \pi r^2 h = \pi \times 6 \times 6 \times 25 = 2827·433 388$
>
> Hence volume of cylinder = 2800 cm³ correct to 2 sig. figs.

ONLINE

Use the 'Significant Number Calculator' to adapt any number to the relevant number of significant figures: www.brightredbooks.net/Applications (try changing the population of the UK from 2011 to 3 sig. figs).

DON'T FORGET

Zeros used to indicate the position of the decimal point are not significant.

THINGS TO DO AND THINK ABOUT

1. Round 6·0092 to three decimal places.

2. Round 5·98 to one decimal place.

3. Round 17·4999 to 2 sig. figs.

4. Round 6786 to 3 sig. figs.

5. Calculate the area of a circle of radius 13·5 cm. Give your answer correct to 2 sig. figs.

ONLINE TEST

Take the 'Significant Figures' test at www.brightredbooks.net/Applications

DECIMALS

Do not use a calculator for any of the examples in this section.

MULTIPLICATION AND DIVISION BY MULTIPLES OF 10, 100 AND 1000

EXAMPLE:

a. $6{\cdot}25 \times 300 = 6{\cdot}25 \times 3 \times 100 = 18{\cdot}75 \times 100 = 1875$

b. $9{\cdot}6 \div 4000 = 9{\cdot}6 \div 1000 \div 4 = 0{\cdot}0096 \div 4 = 0{\cdot}0024$

ORDER OF OPERATIONS

Some expressions involve several operations – for example, addition, subtraction, multiplication, division, as well as brackets, powers of numbers and roots. With these more complicated expressions, the operations must be carried out in the correct order.

You should do the operations in brackets first, before dealing with any orders – that is, powers or roots – and then you should multiply or divide before you add or subtract.

EXAMPLE:

The following formula is used to calculate the cost of electricity in pounds:

Cost in pounds = 25 + number of units × 0·09

Find the cost of 800 units of electricity.

SOLUTION:

$$\begin{aligned}
\text{Cost in pounds} &= 25 + \text{number of units} \times 0{\cdot}09 \\
&= 25 + 800 \times 0{\cdot}09 \\
&= 25 + 8 \times 100 \times 0{\cdot}09 \\
&= 25 + 8 \times 9 \\
&= 25 + 72 \\
&= 97
\end{aligned}$$

Cost of electricity = £97.

EXAMPLE:

A company produces spring water in 500 ml bottles. Each empty bottle weighs 80 g and 1 l of spring water weighs 1 kg.

The bottles are transported to shops in crates that contain 100 bottles. A supermarket has ordered 30 crates of spring water. The delivery van can carry a maximum load of 1·5 tonnes. Can the order be safely delivered in one van load? Use your working to justify your answer.

SOLUTION:

If 1 l (1000 ml) of spring water weighs 1 kg (1000 g), then 500 ml weigh 500 g.

Hence the weight of the bottle plus spring water = 80 g + 500 g = 580 g

Total weight of all the bottles = 30 × 100 × 580 g = 1 740 000 g = 1740 kg

As 1 tonne = 1000 kg, 1740 kg = 1·74 tonnes.

The bottles cannot be delivered in one van load as 1·74 > 1·5.

DECIMALS IN SPORT

We encounter decimals in many sports, from the times of athletes and swimmers, to the scores in diving and gymnastics, to average scores in golf and cricket.

EXAMPLE:

To work out the batting average of a cricketer, divide the total number of runs scored by the cricketer by the number of times he or she has been out. For example, a cricketer with scores of 10, 60 not out, 0 and 50 would have an average of (10 + 60 + 0 + 50) ÷ 3 = 120 ÷ 3 = 40.

In the 2013–14 Ashes series of test matches between Australia and England, the scores of the English batsman Ian Bell were as follows:

	1st innings	2nd innings
1st test	5	32
2nd test	72 not out	6
3rd test	15	60
4th test	27	0
5th test	2	16

Calculate his average for the series (correct to two decimal places).

**SOLUTION:**

Average = (5 + 32 + 72 + 6 + 15 + 60 + 27 + 0 + 2 + 16) ÷ 9 = 235 ÷ 9 = 26·11

Working for division:

$$
\begin{array}{r}
26 \cdot 111 \\
\hline
9\,)\,2\,3^{5}5 \cdot {}^{1}0^{1}0^{1}0
\end{array}
$$

 DON'T FORGET

It is very important that you can divide a number to two decimal places without a calculator. Note that, in this example, three zeros were added after the decimal point because the third number tells you how to round.

ONLINE TEST

Test yourself on decimals at www.brightredbooks.net/ Applications

THINGS TO DO AND THINK ABOUT

1. Work out:

   a. 6·35 × 4000.

   b. 612 ÷ 40.

2. The cost, in pounds, of supplying gas to a property is given by the formula:

   Cost = 25 + 0·12 × number of units

   Calculate the cost of 1300 units of gas.

3. Carly knows she will qualify for the final stages of the club golf championship in August if her average score is less than 72 in the six medal events in July. Her scores are: 70, 69, 77, 74, 71, 69.

   a. Calculate her mean score correct to two decimal places.

   b. Will she qualify for the final stages of the club championship?

FRACTIONS

COMPARING THE SIZE OF FRACTIONS

> **EXAMPLE:**
>
> John and Fiona both receive the same amount of pocket money.
>
> Each week John saves $\frac{2}{3}$ of his pocket money and Fiona saves $\frac{3}{5}$ of her pocket money.
>
> Who has saved the most?
>
> Give a reason for your answer.

> **SOLUTION:**
>
> The two fractions have different denominators. We can only compare their sizes when they have the same denominator. So we must find fractions equivalent to $\frac{2}{3}$ and $\frac{3}{5}$ with the same denominator. This common denominator should be the **least common multiple** (LCM) of 3 and 5. As the LCM of 3 and 5 is 15, we convert both fractions to fifteenths.
>
> $\frac{2}{3} = \frac{\square}{15}$ and $\frac{3}{5} = \frac{\square}{15}$ leading to $\frac{2}{3} = \frac{10}{15}$ and $\frac{3}{5} = \frac{9}{15}$
>
> Now, because $\frac{10}{15}$ is greater than $\frac{9}{15}$, we can see that $\frac{2}{3}$ is greater than $\frac{3}{5}$.
>
> Hence John saves the most because $\frac{2}{3} > \frac{3}{5}$

ADDITION AND SUBTRACTION OF FRACTIONS

You should know how to add simple fractions without using a calculator.

> **EXAMPLE:**
>
> Calculate $\frac{3}{4} + \frac{1}{5}$

> **SOLUTION:**
>
> $\frac{3}{4} + \frac{1}{5} = \frac{3 \times 5}{20} + \frac{1 \times 4}{20} = \frac{15}{20} + \frac{4}{20} = \frac{19}{20}$

Note that we find the LCM of 4 and 5, which is 20 (4 × 5), and then form two equivalent fractions with a denominator of 20.

To subtract fractions, we use a similar approach to addition – that is, we first find the LCM.

> **EXAMPLE:**
>
> Calculate $\frac{3}{4} - \frac{1}{5}$.

> **SOLUTION:**
>
> $\frac{3}{4} - \frac{1}{5} = \frac{3 \times 5}{20} - \frac{1 \times 4}{20} = \frac{15}{20} - \frac{4}{20} = \frac{11}{20}$

> **EXAMPLE:**
>
> Yang spends $\frac{1}{6}$ of his weekly net pay on rent and $\frac{3}{20}$ on transport.
>
> What fraction of his net pay remains?

> **SOLUTION:**
>
> Fraction spent on rent and transport $= \frac{1}{6} + \frac{3}{20} = \frac{10}{60} + \frac{9}{60} = \frac{19}{60}$
>
> Fraction remaining $= 1 - \frac{19}{60} = \frac{60}{60} - \frac{19}{60} = \frac{41}{60}$

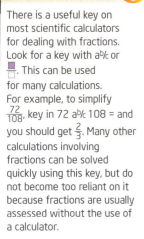

DON'T FORGET

There is a useful key on most scientific calculators for dealing with fractions. Look for a key with a^b/$_c$ or ▦. This can be used for many calculations. For example, to simplify $\frac{72}{108}$, key in 72 a^b/$_c$ 108 = and you should get $\frac{2}{3}$. Many other calculations involving fractions can be solved quickly using this key, but do not become too reliant on it because fractions are usually assessed without the use of a calculator.

MULTIPLICATION OF FRACTIONS

To multiply fractions, multiply the numerators together and the denominators together, then express the resulting fraction in its simplest form.

EXAMPLE:

Calculate $\frac{2}{3} \times \frac{5}{8}$

SOLUTION:

$\frac{2}{3} \times \frac{5}{8} = \frac{2 \times 5}{3 \times 8} = \frac{10}{24} = \frac{5}{12}$

DIVISION OF FRACTIONS

To divide fractions, we leave the first fraction as it is, change the division sign to a multiplication sign, invert the second fraction – that is, turn it upside down – then carry out the calculation in the same way as for the multiplication of fractions.

EXAMPLE:

Calculate $\frac{2}{9} \div \frac{1}{3}$.

SOLUTION:

$\frac{2}{9} \div \frac{1}{3} = \frac{2}{9} \times \frac{3}{1} = \frac{2 \times 3}{9 \times 1} = \frac{6}{9} = \frac{2}{3}$.

EXAMPLE:

Lynne is packing boxes along a shelf in a factory. Each box is $12\frac{1}{2}$ cm long.

a. How many boxes will fit along the shelf if the shelf is 160 cm long?

b. What space will be left over?

SOLUTION:

a. $160 \div 12\frac{1}{2} = \frac{160}{1} \div \frac{25}{2} = \frac{160}{1} \times \frac{2}{25} = \frac{320}{25} = \frac{64}{5} = 12\frac{4}{5}$

Hence 12 boxes would fit on the shelf.

b. Amount of space used up = $12 \times 12\frac{1}{2} = \frac{12}{1} \times \frac{25}{2} = \frac{300}{2} = 150$ cm

Space left over = 10 cm.

THINGS TO DO AND THINK ABOUT

1. In an election for school captain, the two leading contenders were Gillian and Lily. Gillian received $\frac{1}{4}$ of the votes cast. Lily received $\frac{3}{10}$ of the votes cast. Who was elected school captain? Give a reason for your answer.

2. Scott has decided to spend $\frac{2}{7}$ of his take-home pay on entertainment each week. He spends £84 each week on entertainment. What is Scott's weekly take-home pay?

3. Matilda earns £7·20 per hour as a machine operator. How much will she earn if she works for $1\frac{3}{4}$ hours?

PERCENTAGES

You should understand the relationships between fractions, decimal fractions and percentages to be able to choose an efficient route to a solution. In particular, you should know the following relationships and how to use them:

$1\% = \frac{1}{100}; \ 10\% = \frac{1}{10}; \ 20\% = \frac{1}{5}; \ 25\% = \frac{1}{4}; \ 33\frac{1}{3}\% = \frac{1}{3}; \ 50\% = \frac{1}{2}; \ 66\frac{2}{3}\% = \frac{2}{3}; \ 75\% = \frac{3}{4}$

VIDEO LINK

Watch a video on how to calculate percentages mentally at www.brightredbooks.net/Applications

EXAMPLE:

Calculate 75% of £4·80

SOLUTION:

75% of £4·80 = $\frac{3}{4}$ of £4·80 = £4·80 ÷ 4 × 3 = £1·20 × 3 = £3·60

DON'T FORGET

Percent means 'out of 100'.

CONVERTING A PERCENTAGE TO A FRACTION

EXAMPLE:

Convert $37\frac{1}{2}\%$ to:

a. a fraction in its simplest form

b. a decimal fraction.

SOLUTION:

a. Start by expressing $37\frac{1}{2}\%$ as $\frac{37\frac{1}{2}}{100}$.

Now double the numerator and denominator (because there is a 2 on the denominator) and then express the fraction in its simplest form.

Hence $37\frac{1}{2}\% = \frac{37\frac{1}{2}}{100} = \frac{75}{200} = \frac{3}{8}$.

b. $\frac{37\frac{1}{2}}{100} = \frac{37\cdot5}{100} = 0\cdot375$.

EXPRESSING ONE QUANTITY AS A FRACTION OF ANOTHER

To convert a fraction to a percentage, multiply by 100.

EXAMPLE:

A consignment of 750 apples is delivered to a supermarket. Sixty of the apples are damaged. What percentage of the apples is damaged?

SOLUTION:

Percentage damaged = $\frac{60}{750} \times 100\% = \frac{6}{75} \times 100\% = \frac{2}{25} \times 100\% = 8\%$.

EQUIVALENCE OF FRACTIONS, DECIMALS AND PERCENTAGES

We know that fractions, decimals and percentages can be equivalent (think of $\frac{1}{2} = 0.5 = 50\%$). This can prove useful when comparing quantities.

EXAMPLE:

Express $\frac{5}{6}$ as (a) a percentage and (b) a decimal number.

SOLUTION:

a. $\frac{5}{6} = \frac{5}{6} \times 100\% = \frac{500}{6}\% = 83\frac{2}{6}\% = 83\frac{1}{3}\%$.

b. $\frac{5}{6} = 5 \div 6 = 0.833$ (correct to three decimal places).

EXAMPLE:

Express 0·84 as (a) a percentage and (b) a fraction.

SOLUTION:

a. $0.84 = 0.84 \times 100\% = 84\%$.

b. $0.84 = \frac{84}{100} = \frac{21}{25}$.

EXAMPLE:

Sahib scored the following marks in his April tests: Maths, 32 out of 40; English, 79%; History, 37 out of 50; Computing, 40 out of 60; PE, 85%.

Put his marks in order from lowest to highest. You must support your answer with working.

SOLUTION:

Convert all the marks to percentages for comparison.

Maths: $\frac{32}{40} \times 100\% = \frac{4}{5} \times 100\% = 80\%$

English: 79%

History: $\frac{37}{50} \times 100\% = 74\%$

Computing: $\frac{40}{60} \times 100\% = \frac{2}{3} \times 100\% = 66\frac{2}{3}\%$

PE: 85%

Therefore the correct order is: Computing, History, English, Maths, PE.

DON'T FORGET

If you have to compare the size of fractions, decimals and percentages, you should convert them to the same form. As it can be difficult to compare the size of certain fractions, it is usually simplest to convert to a percentage or decimal form.

THINGS TO DO AND THINK ABOUT

You may use a calculator for Q1 only.

1. Robert designs a website to publicise his hotel. In the first month, 450 guests stay in the hotel. In the following month, 567 guests stay in the hotel.

   Calculate the percentage increase in the number of guests.

2. In his exams, Stewart scores the following marks: Chemistry, 48 out of 64; Physics, 38 out of 50; and Biology, 72%.

   He thinks that Chemistry is his best subject. Is he correct? Justify your answer.

ONLINE TEST

Test yourself on percentages at www.brightredbooks.net/ Applications

APPRECIATION

GOING UP!

In mathematics, appreciation refers to a quantity going up in value over a period of time. If money deposited in a bank earns interest over a period of time, we say that the money appreciates. This section considers two methods of finding solutions to problems about appreciation.

EXAMPLE:

Jonathan buys an antique chair for £1500. It is predicted that its value will increase at the rate of 4·5% per year.

What will be the predicted value of the chair in three years? Give your answer to the nearest pound.

SOLUTION:

Method 1: one year at a time

First year: increase in value = 4·5% of £1500 = $\frac{4·5}{100}$ × £1500 = £67·50.

New value = £1500 + £67·50 = £1567·50.

Second year: increase in value = 4·5% of £1567·50 = $\frac{4·5}{100}$ × £1567·50 = £70·5375.

New value = £1567·50 + £70·537 5 = £1638·0375.

Third year: increase in value = 4·5% of £1638·0375 = $\frac{4·5}{100}$ × £1638·0375 = £73·7116875.

New value = £1638·0375 + £73·7116875 = £1711·749188.

Hence predicted value after three years is £1712 (to the nearest pound).

SOLUTION:

Method 2: using a **multiplier**

Multiplier = 100% + 4·5% = 104·5% = 1·045.

Predicted value after three years = £1500 × 1·045 × 1·045 × 1·045 = £1711·749188.

Note: £1500 × 1·045 × 1·045 × 1·045 would normally be written as £1500 × $1·045^3$.

Hence the predicted value after three years is £1712 (to the nearest pound).

Advantages and disadvantages

Many students use Method 1, as they feel confident working out percentages and adding. However, it is time consuming and is unsuitable for periods of time longer than 3 years. The chance of making a mistake increases with each calculation and early rounding could lead to an incorrect final solution, so all figures must be included throughout the calculation. By contrast, Method 2 is quick and efficient and avoids the pitfalls of Method 1. There will be occasions when Method 1 may be suitable, but it is essential that you learn, practise and become efficient in using Method 2.

EXAMPLE:

The population of Newton is 35 000. It is predicted that the population will increase by 6% per year. What will be the predicted population in five years?

Give your answer to the nearest thousand.

SOLUTION:

Predicted population = 35 000 × $1·06^5$ = 46 837·895 22.

Hence predicted population = 47 000 (to the nearest thousand).

DON'T FORGET

You should know how to use your calculator to find $1·045^3$ without having to enter 1·045 × 1·045 × 1·045.

COMPOUND INTEREST

When money is deposited in a bank for periods longer than one year, the interest is usually calculated as compound interest.

EXAMPLE:

Danny has deposited £10 000 in a Platinum Account. It offers a fixed rate of compound interest at 3·1% per annum.

Assuming Danny does not withdraw any money from his account, how much compound interest will he receive over a period of six years?

SOLUTION:

Amount in bank after six years = £10 000 × $1·031^6$ = 12 010·248 45

Amount = £12 010·25 (to the nearest penny)

Hence compound interest received = £12 010·25 − £10 000 = £2010·25.

ANNUAL PERCENTAGE RATE

When borrowing money, many companies charge interest monthly, say 1·125%. If this is then calculated as compound interest, we can find the interest rate for a full year. This is known as the annual percentage rate or APR.

EXAMPLE:

A loan company charges its customers 1·125% per month. Find the APR.

SOLUTION:

No matter how much is borrowed, the APR will be the same. First, find the multiplier (1·011 25), then calculate as compound interest for 12 months to make up a full year.

Hence $1·011 25^{12}$ = 1·143 674 441.

This means that the multiplier for a full year is 1·143 674 441. By subtracting 1, we can see that the annual increase, as a decimal, is 0·143 674 441. We multiply by 100 to find the APR: 0·143 674 441 × 100 = 14·367 444 1%. This would normally be rounded to one decimal place to give 14·4%.

If you borrow money with this company, the monthly interest rate is equivalent to over 14% interest over a full year. It is common for people to borrow money to buy large items such as houses and cars. A 'loan shark' is someone who loans out money, but at a very high rate of interest. When borrowing money, it is important to check that you will be able to pay back the interest and that the APR is not excessively high.

THINGS TO DO AND THINK ABOUT

1. A test on bacteria is carried out in a hospital. The number of bacteria increases at the rate of 0·5% per hour. There are 2000 bacteria at 9 am. How many bacteria will there be at 1 pm? Give your answer correct to three significant figures.

2. A sum of £500 has been deposited in a bank account. How much will be in the bank account after three years if the rate of compound interest is 1·75% per annum?

3. The value of a house increases from £175 000 to £196 000 in one year.

   a. What was the percentage increase?

   b. If the value of the house continues to increase at this rate, what will its value be after a further three years? Give your answer correct to three significant figures.

VIDEO LINK

Watch the 'Compound Interest' clip prepared by a bank in New Zealand at www.brightredbooks.net/Applications

DON'T FORGET

You should investigate Annual Percentage Rate (APR) both online and by checking out adverts which mention APR in newspapers and magazines, paying close attention to the rates of interest.

ONLINE TEST

Take the test 'Appreciation' online at www.brightredbooks.net/Applications

DEPRECIATION

GOING DOWN!

In mathematics, depreciation refers to a quantity that goes down in value over a period of time. When a car decreases in value over a period of time, we say that its value has depreciated. Examples of depreciation are very similar to those on appreciation. We use Method 2 from the previous section (using a multiplier). This time, however, the multiplier is found by subtracting from 100.

EXAMPLE:

A car is valued at £9600. Its value is expected to depreciate by 15% per year.

To the nearest £100, what will the car be worth after three years?

SOLUTION:

Multiplier = 100% − 15% = 85% = 0·85.

Predicted value after three years = £9600 × 0·85³ = £5895·60.

Hence predicted value after three years is £5900 (to the nearest £100).

EXAMPLE:

A factory uses a piece of machinery that costs £55 000 new and will be replaced at the end of the year in which its value falls below half of this initial value. If its value decreases by 20% per year, when should it be replaced?

SOLUTION:

This is slightly different, as we do not know how many years will be involved, so we have to calculate the value of the piece of machinery one year at a time. We can still use a multiplier, however.

Half of initial value when new is £55 000 ÷ 2 = £27 500

Multiplier = 100% − 20% = 80% = 0·8

Value after one year = £55 000 × 0·8 = £44 000

Value after two years = £44 000 × 0·8 = £35 200

Value after three years = £35 200 × 0·8 = £28 160

Value after four years = £28 160 × 0·8 = £22 528

Hence the piece of machinery should be replaced after four years, as £22 528 < £27 500.

> **BEWARE: Some candidates mistakenly say that the machinery should be replaced at the end of the third year because 3 × 20% = 60% and 60% > 50%. It does not work like this and the calculation must be carried out as shown in the example.**

GOING UP AND DOWN!

Values may increase and then decrease, or vice versa. For example, many people buy shares in companies as an investment. The value of shares can go up as well as down. One measure of the performance of shares is the Financial Times Stock Exchange (FTSE) or 'Footsie' index.

contd

DON'T FORGET

Many questions on appreciation and depreciation ask for the answer to be rounded and many students forget this final operation. Always check carefully if rounding is required.

EXAMPLE:

At the start of one week, the FTSE index was 6000.

On Monday, it rose by 3·2%; on Tuesday, it rose by 1·3%; on Wednesday, it fell by 2·1%; on Thursday, it fell by 0·2%; and, on Friday, it rose by 1·9%.

What was the FTSE index at the end of the week?

SOLUTION:

This value can also be calculated using multipliers, although in this case there are five multipliers that must be used one after the other:

FTSE index = 6000 × 1·032 × 1·013 × 0·979 × 0·998 × 1·019 = 6244·933 386

Hence at the end of the week the FTSE index was 6245.

Note: Check all your working carefully, paying particular attention to the five multipliers.

EXAMPLE:

The population of Ashtown is 53 000. The population of Oakville is 40 000. It is predicted that the population of Ashtown will decrease by 5% per year and the population of Oakville will increase by 4% per year.

How many years will it take before the population of Ashtown is less than the population of Oakville?

SOLUTION:

It will be four years before the population of Ashtown is less than Oakville.

Time (years)	Ashtown	Oakville
1	53 000 × 0·95 = 50 350	40 000 × 1·04 = 41 600
2	50 350 × 0·95 = 47 833	41 600 × 1·04 = 43 264
3	47 833 × 0·95 = 45 441	43 264 × 1·04 = 44 995
4	45 441 × 0·95 = 43 169	44 995 × 1·04 = 46 794

 ONLINE

You will find plenty of worked examples to think about and problems to try at www.brightredbooks.net/Applications

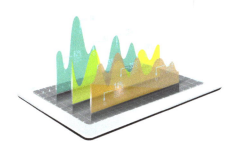

NON-CALCULATOR PROBLEMS

It would be unusual to be asked to solve a problem on appreciation and depreciation without a calculator. If you were asked to do this, the figures would be fairly easy to manipulate.

EXAMPLE:

The value of a laptop decreased from £400 to £320 in one year.

a. What was the percentage decrease?

b. If the value of the laptop continues to decrease at this rate, what will the value be after a further year?

SOLUTION:

a. Actual decrease = £400 − £320 = £80

Percentage decrease = $\frac{80}{400} \times 100 = \frac{1}{5} \times 100 = 20\%$.

b. Decrease in value next year = 20% of £320 = £64

Value after further year = £320 − £64 = £256.

Check that you are comfortable doing these calculations without a calculator.

 ## THINGS TO DO AND THINK ABOUT

A caravan is valued at £15 000. Its value is expected to depreciate by 12% per year. What will the caravan be worth after five years? Give your answer to the nearest £100.

 ONLINE TEST

Take the test 'Depreciation' online at www.brightredbooks.net/Applications

RATIOS

We come across ratios on many occasions. We see ratios in scales on a map, such as 1:100. In certain jobs, there are ratios that should be observed — for example, the ratio of adults to children in a nursery group should be 1:13 for children older than three years. If we want to use a recipe designed to serve four people and we have seven people arriving for a meal, the quantities will have to be increased in the ratio 7:4.

ONLINE

Go to www.brightredbooks.net/ Applications for a revision of ratios.

SHARING AN AMOUNT IN A RATIO

EXAMPLE:

Peter has £2·50, Paul has £3·50 and Mary has £4. They put their money together to buy some lottery tickets. They win £160.

a. Write down the amounts spent by Peter, Paul and Mary as a ratio in its simplest form.

b. Share their winnings in the ratio from part (a).

SOLUTION:

a. Express all amounts in the same units (pence).

Ratio = 250:350:400 = 5:7:8

b. Add the numbers in the ratio, so that 5 + 7 + 8 = 20

Peter receives $\frac{5}{20}$ of £160 = £40.

Paul receives $\frac{7}{20}$ of £160 = £56.

Mary receives $\frac{8}{20}$ of £160 = £64.

DON'T FORGET

In questions where you are sharing an amount in a ratio, add up your answers as a check: £40 + £56 + £64 = £160.

EXAMPLE:

A science centre is arranging visits for parents, teachers and pupils. The ratio of parents to teachers to pupils must be 1:2:10.

a. 120 pupils apply to go on the visit. How many teachers must accompany them?

b. The science centre gives the school 150 tickets for the exhibition.

What is the maximum number of pupils who can attend the exhibition?

SOLUTION:

a. The ratio of teachers to pupils is 2:10 – that is, one teacher for every five pupils. Hence, because 120 ÷ 5 = 24, 24 teachers should accompany 120 pupils.

b. Add the numbers in the ratio, so that 1 + 2 + 10 = 13

Number of groups of 13 = 150 ÷ 13 = 11·538 = 11

Number of pupils = 11 × 10 = 110

RATIOS AND RECIPES

EXAMPLE:

Sean looks up a recipe to make marmalade.
The ratio of oranges to sugar in the marmalade is 3:2.

To produce a healthier recipe, Sean decides to:

- increase the orange content by 10%.
- decrease the sugar content by 10%.

Calculate the new ratio of oranges to sugar.
Give your answer as a ratio in its simplest form.

SOLUTION:

In original recipe, there are three parts orange to two parts sugar.

In new recipe, parts of orange = 3 × 1·1 = 3·3

In new recipe, parts of sugar = 2 × 0·9 = 1·8

New ratio of oranges to sugar – 3·3:1·8 – 33:18 – 11:6.

EXAMPLE:

The owner of a delicatessen blends speciality salts.

She blends Himalayan salt and seaweed-infused salt in the ratio 2:3.

Himalayan salt costs her £10 per kilogram.

Seaweed-infused salt costs her £20 per kilogram.

a. Find the cost of 1 kg of the blend.

b. She sells the blend at £28 per kilogram.

   Express the profit that the owner of the delicatessen makes as a percentage of the cost price.

c. As a result of rising expenses, she wishes to make a 125% profit on the blend.

   What should she sell it for now?

SOLUTION:

a. Cost of 5 kg of blend = (2 × £10) + (3 × £20) = £80

   Cost per kilogram = £80 ÷ 5 = £16.

b. Actual profit per kilogram = £28 – £16 = £12

   Percentage profit = $\frac{12}{16}$ × 100% = 75%.

c. Profit = $\frac{125}{100}$ × £16 = £20

   Selling price = £20 + £16 = £36.

 ## THINGS TO DO AND THINK ABOUT

Ebenezer has left his three grandchildren £600 in his will. The amount will be divided between Robert, Catriona and Louis in the ratio 5:4:3. How much will each grandchild get?

MAPS AND PLANS

REMINDERS

This section is closely related to the sections on Scale drawings and Navigation in Part 2, so it would be useful to look back at pages 44–47.

EXAMPLE:

The sitting room in Lucy's house is rectangular, 6 m long and 3·6 m wide.

She has drawn a plan of the sitting room with length 30 cm.

What is the width of the plan?

SOLUTION:

Actual distance (m)	Distance on plan (cm)
6	30
3·6	$\frac{3·6}{6} \times 30 = 18$

The width of the plan is 18 cm.

MAP SKILLS

EXAMPLE:

Using the map of the Isle of Skye, calculate the distance from Armadale to Boreraig (as the crow flies). Give your answer to the nearest 5 km.

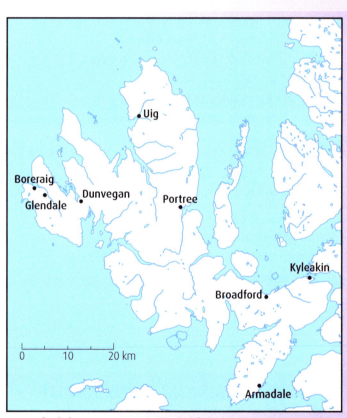

SOLUTION:

Use the scale on the map to find that 2·5 cm represents 20 km.

On the map, the distance from Armadale to Boreraig = 8 cm (approximately). Hence:

Centimetres	Kilometres
2·5	20
8	$\frac{8}{2·5} \times 20 = 64$

The distance from Armadale to Boreraig is 65 km (to the nearest 5 km).

PLOTTING POINTS ON A MAP

In the earlier section on scale drawings, we looked at three-figure bearings and how to find the position of a port or ship on a map. The following example shows how to plot the position of a point on a map relative to other places.

EXAMPLE:

A ship is in distress in the North Sea. It is on a bearing of 125° from Aberdeen and on a bearing of 230° from Stavanger in Norway. Show the position of the ship on the map.

SOLUTION:

Draw a north line at Aberdeen and measure 125° clockwise, then draw a line. Draw a north line at Stavanger and measure 230° clockwise (or 130° anticlockwise), then draw a line. The intersection of the two lines shows the position of the ship in distress. The scale on the map is not required for this, although it could be used to calculate the distances from Aberdeen and Stavanger to the ship.

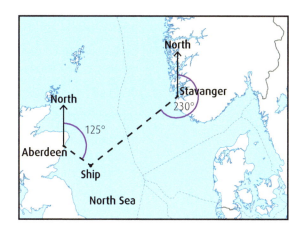

PLANS

EXAMPLE:

The diagram shows a plan of a football pitch. The plan is drawn to scale.

The length of the football pitch is 105 m. Calculate its breadth.

SOLUTION:

The length of the football pitch is measured as 12 cm.

The breadth of the football pitch is measured as 8 cm.

Now calculate the breadth using proportion.

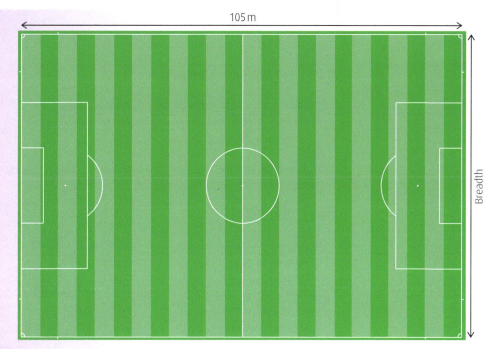

Centimetres	Metres
12	105
8	$\frac{8}{12} \times 105 = 70$

Hence the breadth of the football pitch is 70 m.

THINGS TO DO AND THINK ABOUT

Using the map of the Isle of Skye, calculate the distance from Dunvegan to Kyleakin (as the crow flies). Give your answer to the nearest 5 km.

ONLINE TEST

Test yourself on maps and plans at www.brightredbooks.net/Applications

DISTANCE, SPEED AND TIME

Problems involving distance, speed and time can be solved using simple, easy to remember formulae.

CALCULATING DISTANCE

To calculate the distance of a journey, we multiply the average speed for the journey by the time taken. This can be written as a formula:

Distance = speed × time ($D = S \times T$)

EXAMPLE:

Darren is planning to drive from Glasgow to Inverness, a distance of 171 miles. He must arrive there for a meeting at 1000 hours. He leaves home at 0615 and drives to his destination in Inverness at an average speed of 48 miles per hour (mph).

Will he arrive in Inverness in time for the start of the meeting? Justify your answer.

SOLUTION:

Time interval from 0615 to 1000 = 3 hours 45 minutes = 3·75 hours

Now find the distance travelled at 48 mph using the formula $D = S \times T$.

$D = S \times T = 48 \times 3 \cdot 75 = 180$

Darren will arrive on time as 180 > 171.

DON'T FORGET

Convert times in minutes to a decimal fraction of an hour when speeds are given in mph or km/h by dividing the number of minutes by 60 – for example, 3 hours 45 minutes = 3·75 hours as 45 ÷ 60 = 0·75.

VIDEO LINK

Watch the 'Speed = distance over time' song at www.brightredbooks.net/ Applications

CALCULATING SPEED

To calculate the average speed of a journey, we divide the distance travelled on the journey by the time taken. This can be written as a formula:

Speed = distance ÷ time $\left(S = \frac{D}{T}\right)$

Common units of speed include miles per hour (mph), kilometres per hour (km/h) and metres per second (m/s).

EXAMPLE:

The circuit for the Monaco Grand Prix is 3·337 km long. The drivers complete 78 laps of the track. In 2015, Nico Rosberg won the race in a time of 1 hour 49 minutes 18·420 seconds. Calculate his average speed in kilometres per hour correct to three decimal places.

SOLUTION:

Distance travelled = 78 × 3·337 = 260·286 km

49 minutes 18·420 seconds = 49·307 minutes (as 18·420 ÷ 60 = 0·307)

1 hour 49·307 minutes = 1·82178 hours (as 49·307 ÷ 60 = 0·82178)

$S = \frac{D}{T} = \frac{260 \cdot 286}{1 \cdot 82178} = 142 \cdot 874$ km/h (correct to three decimal places).

CALCULATING TIME

To calculate the time taken for a journey, we divide the distance travelled on the journey by the average speed. This can be written as a formula:

Time = distance ÷ speed $\left(T = \frac{D}{S}\right)$

EXAMPLE:

Mohammed must reach the airport at 1410 to check in for a flight. His home is 27 miles from the airport. He expects to drive at an average speed of 36 miles per hour for the journey. What is the latest time that Mohammed should leave his home.

Justify your answer.

SOLUTION:

$T = \frac{D}{S} = \frac{27}{36} = 0{\cdot}75$ hours

0·75 hours = 0·75 × 60 minutes = 45 minutes.

He should leave 45 minutes before 1410 – that is, at 1325.

THE 'DISTANCE, SPEED, TIME' TRIANGLE

We have now used three formulae for calculating distance, speed and time.

$$D = S \times T$$

$$S = \frac{D}{T}$$

$$T = \frac{D}{S}$$

An easy way to remember the formulae is to use the 'distance, speed, time' triangle.

 Distance = Speed x Time

 Time = $\dfrac{\text{Distance}}{\text{Speed}}$

 Speed = $\dfrac{\text{Distance}}{\text{Time}}$

 **DON'T FORGET**

To help you memorise what goes where in the triangle, think of the word DiSTance.

To use the triangle, cover up the quantity you wish to find with your finger. What you still see shows you the correct formula and tells you whether to multiply or divide.

 ## THINGS TO DO AND THINK ABOUT

Anne Marie drove from Edinburgh to Aberdeen, 125 miles away. She left Edinburgh at 6:45 pm. She arrived in Aberdeen at 9:15 pm.

a. How long did Anne Marie's journey take?

b. Calculate her average speed in miles per hour for the journey.

 ONLINE TEST

Test yourself on distance, speed and time at www.brightredbooks.net/Applications

READING FROM A SCALE

There is a variety of measuring instruments in every household that require us to read scales. There are clocks and timers, scales for weighing and jugs for measuring liquids in the kitchen. A boiler may have timers and temperature settings. A barometer measures atmospheric pressure and may include a thermometer. Many bathrooms have a set of scales. Get into a car and you will see a distance gauge, a speedometer and a petrol gauge. It is vital that you can read the scales on such instruments.

VIDEO LINK

Get some good advice on reading a scale at www.brightredbooks.net/ Applications

EXAMPLE:

The photo shows a thermometer that measures temperature in both degrees Celsius and degrees Fahrenheit.

Estimate

a. the temperature in degrees Fahrenheit and

b. the temperature in degrees Celsius.

SOLUTION:

a. 26°F or 27°F

b. −3°C.

DON'T FORGET

When reading scales, you must find the value of each division on the scale. A simple way to do this is to find the first marked number on the scale after zero and then to divide by the number of spaces between 0 and that number. In the scale for weighing shown here, we divided 100 by 10 to find that each division represented 10 g.

EXAMPLE:

Delia needs 8 ounces of tomatoes for a recipe. She places two tomatoes on the scales.

Does she have enough tomatoes for the recipe?

She knows that 1 ounce = 28 grams (approximately).

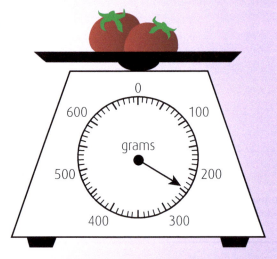

SOLUTION:

There are 10 divisions (spaces) between 0 and 100, so each division = 10 g. Therefore the tomatoes weigh 240 g.

8 ounces = 8 × 28 g = 224 g

Delia does have enough tomatoes as 240 > 224.

EXAMPLE:

Mark recently filled up his petrol tank. The tank holds 56l when full.

He looks at the petrol gauge in his car.

a. What is the reason for the red marking at the left-hand end of the gauge?

b. What fraction of the tank contains petrol?

c. How much petrol has Mark used so far?

SOLUTION:

a. It tells you when petrol is low and warns you to add petrol to the tank.

b. $\frac{7}{8}$.

c. $\frac{1}{8}$ of 56l = 7l.

EXAMPLE:

Jan takes her car from the UK to France for a family holiday. As she drives along a French road, she looks at her speedometer.

She notices road signs indicating that the speed limit is 110 km/h.

She knows that 5 miles is equal to 8 km.

Is Jan exceeding the speed limit? Justify your answer.

SOLUTION:

Jan's speed is 64 mph.

Miles	Kilometres
5	8
64	$\frac{64}{5} \times 8 = 102 \cdot 4$

Jan is not exceeding the speed limit as 102·4 < 110 km/h.

THINGS TO DO AND THINK ABOUT

Orange juice has been poured into a measuring jug. How many millilitres of orange juice should be added to fill the jug to the 1·5l mark?

1·5 litres

ONLINE TEST

Test yourself on reading from a scale at www.brightredbooks.net/Applications

STEM AND LEAF DIAGRAMS

A stem and leaf diagram presents data in a clear and ordered way. It has the added bonus that it shows every member in a data set.

EXAMPLE:

The stem and leaf diagram shows the heights, to the nearest centimetre, of a group of boys:

```
15 │ 9
16 │ 2  5  7  8  9
17 │ 3  5  6  7  8  9
18 │ 0  4  5  6
```

(n = 16)   15 │ 9   represents 159 centimetres

What is the probability that a boy, chosen at random, has a height of more than 180 cm?

SOLUTION:

There are three students with a height of more than 180 cm (184, 185 and 186 cm), therefore the probability is $\frac{3}{16}$.

A back-to-back stem and leaf diagram can be used to compare two sets of data. Look back to the example on p. 33 for a reminder of a back-to-back stem and leaf diagram. We look at a longer case study in this section.

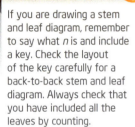

DON'T FORGET

If you are drawing a stem and leaf diagram, remember to say what n is and include a key. Check the layout of the key carefully for a back-to-back stem and leaf diagram. Always check that you have included all the leaves by counting.

CASE STUDY

EXAMPLE:

The results of a group of students for their Chemistry exam (out of 50) are shown in a back-to-back stem and leaf diagram:

Boys   Girls

```
              5  2 │ 1 │ 8
        7  6  5  1 │ 2 │ 2  6
     9  8  6  2  1 │ 3 │ 2  4  5  6  7  8  9
        7  5  3  0 │ 4 │ 1  3  3  3  5
```

n = 15 n = 15

```
                   │ 1 │ 8      means 18
                 2 │ 1 │        means 12
```

(a) A boxplot is drawn to represent one set of data.

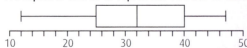

```
10        20        30        40        50
```

Does the boxplot represent the boys' data or the girls' data?

Give a reason for your answer.

(b) For the other set of data, find:

   (i) the median   (ii) the lower quartile   (iii) the upper quartile.

(c) Use the answers to part (b) to draw a boxplot for the other data set.

(d) Make two appropriate comments about the distribution of the data in the two sets.

contd

VIDEO LINK

Watch a video on drawing stem and leaf diagrams at www.brightredbooks.net/Applications

SOLUTION:

(a) The boxplot represents the boys' data. There are numerous reasons you could give. One of the simplest is that it shows the lowest mark to be 12. This matches the boys' data.

(b) As $n = 15$, the position of median = $(15 + 1) \div 2 = 8$. This means the median is the eighth value. As the numbers are ordered, the median for the girls is 37. The lower quartile is the median of the numbers in the lower half (18, 22, 26, 32, 34, 35 and 36) and is therefore 32. The upper quartile is the median of the numbers in the upper half (38, 39, 41, 43, 43, 43 and 45) and is therefore 43.

Hence the solutions are: (i) 37; (ii) 32; and (iii) 43.

(c) The lowest value = 18; the highest value = 45.

Therefore boxplot for girls:

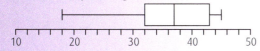

(d) The girls performed better than the boys because their median value (37) is higher than that of the boys (32). The boxplot for the girls is not as spread out as the boxplot for the boys. This indicates that the results for the girls are more consistent.

 ONLINE TEST

Test yourself on stem and leaf diagrams at www.brightredbooks.net/Applications

 ## THINGS TO DO AND THINK ABOUT

Rovers and United are due to meet in a cup tie. The ages of the players in the squad for Rovers are:

32	33	28	27	23	30
24	25	29	33	31	23
19	26	30	34	22	40

The ages of the players in the squad for United are:

22	24	28	26	25	23
32	25	21	20	17	19
18	26	24	29	22	34

a. Draw an appropriate statistical diagram to compare the ages of the two squads.

b. Make one appropriate comment about the distribution of the ages in the two squads.

LINE GRAPHS

Line graphs provide a useful way of illustrating information that changes, often over time. Examples could be the temperature of a patient in a hospital measured every few hours, the population of a country measured each year, the daily change in the FTSE index on the stock exchange or the weekly position of a tennis player in the world rankings.

INTERPRETING A LINE GRAPH

EXAMPLE:

Margo is in hospital with a high temperature. After she has been given some medication, a nurse checks her temperature every two hours and records it on a line graph, part of which is shown here.

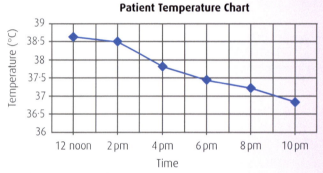

a. What was Margo's temperature at 12 noon?

b. What is the trend of the graph?

c. Normal body temperature is 37°C. When does Margo's temperature reach this value?

SOLUTION:

a. 38·6°C.

b. Downwards.

c. Around 9 pm.

Note: In the part of the graph shown here, the vertical scale starts at 36°C. It is always desirable to start the vertical scale in a line graph at 0. However, this can lead to a large area of empty space. To get around this, it is common to insert a break in the axis shown by a little zig-zag (see the diagram at the start of this section).

VIDEO LINK

Watch a video on line graphs at www.brightredbooks.net/Applications

INTERPOLATION AND EXTRAPOLATION

Interpolation is the process of estimating the value of a variable at a point in between values that are already known. We used interpolation in the previous example to estimate the time (around 9 pm) when the Margo's temperature was 37°C.

Extrapolation is the process of estimating the value of a variable that lies outside a known range of values. It is often used to predict future events.

contd

EXAMPLE:

The weight, in kilograms, of baby John was recorded every two weeks after his birth. The results are illustrated as a line graph.

Estimate the weight of baby John after 20 weeks.

John's Weight

SOLUTION:

The trend of the graph is upwards and the increase is fairly constant. Therefore, by extending the line, you can extrapolate the graph to estimate that John's weight should be almost 6 kg after 20 weeks.

DRAWING A LINE GRAPH

If you are asked to draw a line graph, observe the following rules:

1. Give the graph a title.

2. Use the horizontal axis for time.

3. Label both axes and choose sensible, easy to read scales.

4. Start the vertical axis at 0. If this leads to a lot of unused space, insert a break in the axis shown by a little zig-zag.

5. Plot the points carefully and join them in order from left to right.

6. Do not join the ends of the graph to the axes.

DON'T FORGET

The time scale should always appear on the horizontal axis on a line graph.

EXAMPLE:

The average monthly rainfall, in millimetres, was recorded in a resort on the Costa del Sol in Spain. The results are shown in the table.

Month	J	F	M	A	M	J	J	A	S	O	N	D
Rainfall (mm)	83	75	59	40	23	13	2	5	15	55	115	98

Illustrate these data in a line graph.

SOLUTION:

Costa del Sol – Rainfall

THINGS TO DO AND THINK ABOUT

The average temperature, in degrees Celsius, was recorded in a resort on the Costa del Sol in Spain. The results are shown in the table. Illustrate these data in a line graph.

Month	J	F	M	A	M	J	J	A	S	O	N	D
Temperature (°C)	12	13	14	16	19	22	25	26	23	19	15	13

ONLINE TEST

Test yourself on line graphs at www.brightredbooks.net/Applications

INFORMATION FROM TABLES AND GRAPHS 1

DISTANCE–TIME GRAPHS

EXAMPLE:

Jim's journey from his workplace in Dundee back to his home is illustrated in the following distance-time graph.

a. How far is it from Jim's workplace to his home?

b. Jim had a puncture on the way home. How long did it take him to repair it?

c. Calculate Jim's speed before the puncture in miles per hour?

d. Calculate Jim's speed after the puncture in miles per hour?

SOLUTION:

a. 25 miles.

b. 20 minutes.

c. Before the puncture, he drove 15 miles in 25 minutes.

   25 minutes = 25 ÷ 60 = 0·416666 hours

   $S = \frac{D}{T} = \frac{15}{0·416666} = 36$, so Jim's speed was 36 mph before the puncture.

d. After the puncture, he drove 10 miles in 20 minutes.

   20 minutes = 20 ÷ 60 = 0·333333 hours

   $S = \frac{D}{T} = \frac{10}{0·333333} = 30$, so Jim's speed was 30 mph after the puncture.

Note: Parts (c) and (d) could also be calculated using proportion. In particular, note that part (d) can be found easily by trebling both quantities.

VIDEO LINK

Watch a video on drawing a time–distance graph at www.brightredbooks.net/Applications

EXTRACTING INFORMATION FROM A TABLE

We are often required to look up information presented in tables, including timetables, tables of prices and tables in holiday brochures. When public transport is used, tables with details of prices can be complicated.

EXAMPLE:

The table shows the rates charged by a car ferry company for journeys between Ardrossan and Brodick valid from 3 April to 25 October 2015.

FARES: ARDROSSAN - BRODICK

All tickets must be purchased prior to boarding the vessel.			Single	Return
🚶	Driver/Passenger Children under 5 travel free, children aged 5–15 travel for half the adult fare.		£3.75	£7.50
🚗	Car or 4x4 (excludes driver)		£15.10	£30.20
🚐	Motorhome (excludes driver)	up to 6m	£15.10	£30.20
		up to 8m	£22.65	£45.30
		up to 10m	£30.20	£60.40
🚙	Caravan, Boat/Baggage trailer	up to 2.5m	£7.55	£15.10
		up to 6m	£15.10	£30.20
		up to 8m	£22.65	£45.30
🏍	Motorcycle		£7.55	£15.10
🚲	Pedal cycles (restricted numbers)		Free	Free
Light goods vehicle exceeding 6 metres in length, or 3.5 tonnes, or 3 metres in height, or 2.3 metres in width are charged at the commercial vehicle rate.				
Pier dues are payable to third parties for this route and have been included in the fare shown.				

a. For how many days of the year in 2015 was this timetable valid?

b. Mr and Mrs Murray have two children aged four and nine years. The family travelled from Ardrossan to Brodick in a motorhome of length 6·30m on 8 August and returned on 22 August. What was the total cost of their journey?

SOLUTION:

a. Count 28 days in April as April has 30 days and the only dates on which the timetable was not valid were 1 and 2 April. Add 31 days for May, 30 days for June, 31 days for July, 31 days for August, 30 days for September and then add on 25 for the first 25 days in October. Hence the answer is 28 + 31 + 30 + 31 +31 + 30 + 25 = 206 days.

b. Adult fare = £7·50 each; four-year-old child was free; fare for the nine-year-old child was one-half of £7·50 = £3·75; fare for motorhome = £45·30.

   Total cost = (2 × £7·50) + £3·75 + £45·30 = £64·05.

THINGS TO DO AND THINK ABOUT

Mr and Mrs Khan have three children aged five, nine and 16 years. The family travelled from Ardrossan to Brodick on 15 August and returned on 18 August in a car of length 4·60m. What was the total cost of their journey?

DON'T FORGET

Always check the notes at the foot of tables as there may be additional information that can affect the outcome – for example, the prices for children.

ONLINE TEST

Test yourself on information from graphs and tables at www.brightredbooks.net/ Applications

INFORMATION FROM TABLES AND GRAPHS 2

INFORMATION FROM A TABLE

EXAMPLE:

The cost of sending a letter in the UK depends on the size and weight of the letter. The table shows the various charges.

Format	Weight (g)	Cost	
		1st class mail	2nd class mail
Letter	0–100	63 pence	54 pence
Large letter	0–100	95 pence	74 pence
	101–250	£1·26	£1·19
	251–500	£1·68	£1·51
	501–750	£2·42	£2·05

Moira sends a small letter weighing 50 g by second class mail, a large letter weighing 180 g by second class mail and a large letter weighing 515 g by first class mail.

Calculate the total cost.

SOLUTION:

Cost of small letter weighing 50 g = 54 pence

Cost of large letter weighing 180 g = £1·19

Cost of large letter weighing 515 g = £2·42

Total cost = £0·54 + £1·19 + £2·42 = £4·15.

COMPOUND BAR GRAPHS

A **compound bar graph** is an extension of an ordinary bar graph that compares two or more quantities at the same time.

EXAMPLE:

The diagram shows the expenditure of a manufacturing company on its three main raw materials from 2012 to 2015.

a. How much was spent on material A in 2012?

b. Which material increased its proportion of the annual expenditure over the four-year period?

c. What was the total expenditure on material B over the four-year period?

d. What percentage of the total expenditure for 2012 was taken up by material C?

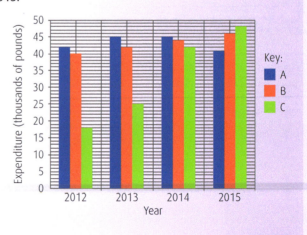

contd

SOLUTION:

a. £42 000.

b. Material C.

c. £40 000 + £42 000 + £44 000 + £46 000 = £172 000.

d. Total expenditure in 2012 = £42 000 + £40 000 + £18 000 = £100 000

Percentage of total taken up by material C = 18%.

DON'T FORGET

When you use a ready reckoner, the value you are looking for may not appear in the table, but it can always be calculated using direct proportion.

READY RECKONERS

A **ready reckoner** is a table of numbers used to ease calculations. They are often used for the conversion of units – for example, between metric and imperial units.

EXAMPLE:

Body mass index (BMI) is used to check whether the weight of a person is healthy. The formula for calculating BMI is:

$$BMI = \frac{\text{weight in kilograms}}{(\text{height in metres})^2}$$

a. Calculate the BMI of Donald, who is 5 feet 8 inches tall and weighs 182·5 pounds.

b. A person with a BMI between 18·5 and 25 is said to have a healthy weight. How many pounds should Donald lose to achieve a healthy weight?

SOLUTION:

a. 5 feet 8 inches = 172·72 cm = 1·7272 m

$182·5$ pounds $= \frac{81·65 + 83·91}{2} = 82·78$ kg

Donald's BMI $= \frac{\text{weight in kilograms}}{(\text{height in metres})^2} = \frac{82·78}{1·7272^2} = 27·75.$

b. For Donald to achieve a healthy BMI, his weight divided by $(1·7272)^2$ should equal 25 at most. Hence:

$$BMI = \frac{\text{weight in kilograms}}{(\text{height in metres})^2} = \frac{\text{weight in kilograms}}{(1·7272)^2} = 25$$

Hence weight in kilograms = $25 \times (1·7272)^2 = 74·58$

So Donald should lose $(82·78 - 74·58)$ kg = 8·2 kg.

Now choose an entry from the ready reckoner to convert to pounds.

Kilograms	Pounds
61·23	135
8·2	$\frac{8·2}{61·23} \times 135 = 18·1$

Therefore Donald should lose at least 18 pounds to reach a healthy weight.

Standard/Metric Conversion Chart			
Feet Inches	**Centimetres**	**Pounds**	**Kilograms**
4 feet 11 inches	149·86	135·00	61·23
5 feet 0 inches	152·40	140·00	63·50
5 feet 1 inch	154·94	145·00	65·77
5 feet 2 inches	157·48	150·00	68·04
5 feet 3 inches	160·02	155·00	70·31
5 feet 4 inches	162·56	160·00	72·57
5 feet 5 inches	165·10	165·00	74·84
5 feet 6 inches	167·64	170·00	77·11
5 feet 7 inches	170·18	175·00	79·38
5 feet 8 inches	172·72	180·00	81·65
5 feet 9 inches	175·26	185·00	83·91
5 feet 10 inches	177·80	190·00	86·18
5 feet 11 inches	180·34	195·00	88·45
6 feet 0 inches	182·88	200·00	90·72
6 feet 1 inch	185·42	205·00	92·99
6 feet 2 inches	187·96	210·00	95·25
6 feet 3 inches	190·50	215·00	97·52
6 feet 4 inches	193·04	220·00	99·79
6 feet 5 inches	195·58	225·00	102·06
6 feet 6 inches	198·12	230·00	104·33
6 feet 7 inches	200·66	235·00	106·59
6 feet 8 inches	203·20	240·00	108·86

THINGS TO DO AND THINK ABOUT

a. Calculate the BMI of Louise, who is 5 feet 4 inches tall and weighs 160 pounds.

b. A person with a BMI between 18·5 and 25 is said to have a healthy weight. How many pounds should Louise lose to achieve a healthy weight?

EXTRA SECTIONS

CASE STUDY 1

GOING ON HOLIDAY

EXAMPLE:

The Smith family are going on holiday to the Costa Brava in Spain. The week before their departure, they note the temperature in the region in degrees Fahrenheit:
80, 82, 79, 77, 79, 83 and 80.

a. Calculate:
   (i)   the mean;
   (ii) the standard deviation of the temperatures.

b. In the same week, the mean temperature on the Costa del Sol in Spain is 86°F and the standard deviation is 0·37°F. Make two appropriate comments about the temperatures in the two regions.

c. On the day of their departure to Spain, the family arrive in a long queue at the airport check-in. Mr Smith notes that it takes 8 minutes for the first 12 people to check in. If the queue continues to move at this rate and there are 54 people ahead of the Smith family in the queue, how long will it be until they can check in?

d. When the family arrives on the Costa del Sol, they decide to go on some trips and calculate the cost of their favourite ones. The family wants to go on three different trips and spend a maximum of 300 euros. Find all the possible options and their costs.

Trips	Cost (Euros)
Trip to Andorra	140
Pirate show	124
Trip to Barcelona	79
Mediaeval banquet	94
Market trip	75

DON'T FORGET

A case study is a longer question of related parts or parts based on a common theme. This case study would be worth around 15 marks in an assessment. As one long question can be intimidating, treat it as a series of short independent questions.

e. While on holiday, Mrs Smith receives a text to her mobile phone informing her that calls to the UK cost 16·5 pence per minute (4·8 pence per minute to receive) and texts cost 4·3 pence to send (free to receive). During the holiday, Mrs Smith made three phone calls to the UK lasting five, six and ten minutes. She received one call from the UK lasting eight minutes. She sent 32 texts and received 40 texts. Mrs Smith had £20·83 credit on her phone when she arrived in the Costa del Sol. How much credit did she have at the end of the holiday?

contd

SOLUTION:

a. (i) Mean = (80 + 82 + 79 + 77 + 79 + 83 + 80) ÷ 7 = 560 ÷ 7 = 80°F.

   (ii) Method 1

x	$x - \bar{x}$	$(x - \bar{x})^2$
80	80 − 80 = 0	0
82	82 − 80 = 2	4
79	79 − 80 = −1	1
77	77 − 80 = −3	9
79	79 − 80 = −1	1
83	83 − 80 = 3	9
80	80 − 80 = 0	0
		Total = 24

Method 2

x	x^2
80	6400
82	6724
79	6241
77	5929
79	6241
83	6889
80	6400
Total = 560	Total = 44 824

$$s = \sqrt{\frac{\Sigma(x - \bar{x})^2}{n - 1}} = \sqrt{\frac{24}{7 - 1}} = \sqrt{\frac{24}{6}} = \sqrt{4} = 2 \text{ or}$$

$$s = \sqrt{\frac{\Sigma x^2 - (\Sigma x)^2/n}{n - 1}} = \sqrt{\frac{44\,824 - 560^2 \div 7}{7 - 1}} = \sqrt{\frac{24}{6}} = \sqrt{4} = 2.$$

b On average, it was warmer on the Costa del Sol as the mean temperature was greater there (86 > 80). The temperatures were more consistent on the Costa del Sol as the standard deviation of the temperature was lower there (0·37 < 2).

c. The family will have another 36 minutes to wait.

Number of people	Minutes
12	8
54	$\frac{54}{12} \times 8 = 36$

d. There are ten distinct ways of choosing three options from five. Start with the most expensive (trip to Andorra) and check the total with two other trips, then continue until all the possible selections are found:

Andorra + Barcelona + market = €140 + €79 + €75 = €294

Pirates + Barcelona + banquet = €124 + €79 + €94 = €297

Pirates + Barcelona + market = €124 + €79 + €75 = €278

Pirates + banquet + market = €124 + €94 + €75 = €293

Barcelona + banquet + market = €79 + €94 + €75 = €248.

e. Cost of phoning UK = (5 + 6 + 10) × 16·5p = 346·5p = £3·47

Cost of receiving calls = 8 × 4·8p = 38·4p = £0·38

Cost of sending texts = 32 × 4·3p = 137·6p = £1·38

Credit left on phone = £20·83 − £3·47 − £0·38 − £1·38 = £15·60.

CASE STUDY 2

IN THE GARDEN

Alan is designing a symmetrical garden with the dimensions shown in the diagram. He plans to include a circular ornamental pond surrounded by grass.

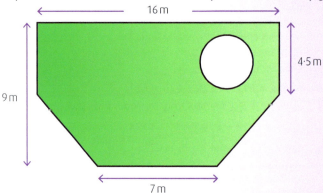

Alan will sow grass seed in the garden. Before he can do that, he must find the area of the lawn.

a. If the diameter of the ornamental pond is 3 m, calculate the area of the lawn.

b. Alan has decided to have a high-quality fence erected around the perimeter of the garden. He makes enquiries regarding the cost and finds the following information:

Quality Gates and Fences
Fencing is sold in sections of 10 ft which can be cut if required   1 ft of high-quality fencing costs £5   The cost of labour is £16 per hour   The expected rate of work is that 8 ft of fencing will be erected per hour   Costs for labour will be rounded up to the next complete hour   There will be a one-off cost of £50 for the use of specialist equipment

Alan knows that 1 m = 39·375 inches and 1 foot = 12 inches.

Calculate the cost of erecting the fence around the garden.

c. The cost of the pond is directly proportional to the square of the diameter. Alan knows that it costs £4800 to install a pond with a diameter of 4 m. Calculate the total cost of installing Alan's pond and fence.

contd

SOLUTION:

a. The simplest method is to calculate the area of the large rectangle surrounding the garden and then subtract the area of two congruent triangles.

Check that each triangle has base 4·5 m [(16 − 7) ÷ 2] and height 4·5 m (9 − 4·5).

Area of surrounding rectangle: $A = lb = 16 \times 9 = 144\,m^2$

Area of each triangle: $A = \frac{1}{2}bh = \frac{1}{2} \times 4·5 \times 4·5 = 10·125\,m^2$

Area of both triangles = $2 \times 10·125 = 20·25\,m^2$

Area of garden = $(144 − 20·25)\,m^2 = 123·75\,m^2$

Area of pond: $A = \pi r^2 = \pi \times 1·5^2 = 7·069\,m^2$

Area of lawn = $(123·75 − 7·069)\,m^2 = 116·681\,m^2 = 117\,m^2$ (to the nearest square metre).

b. We must use the Pythagoras' theorem to calculate the length of the sloping edges of the garden.

$AB^2 = 4·5^2 + 4·5^2 = 20·25 + 20·25 = 40·5$

$\Rightarrow AB = \sqrt{40·5} = 6·36\,m$ (correct to two decimal places).

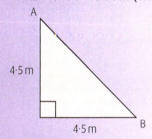

Perimeter of garden = $(16 + 4·5 + 6·36 + 7 + 6·36 + 4·5)\,m = 44·72\,m$

The perimeter must now be converted into feet in order to proceed.

44·72 m = 44·72 × 39·375 inches = 1760·85 inches

1760·85 inches = 1760·85 ÷ 12 feet = 146·7 feet

Now as fencing is sold in multiples of 10 ft, 150 ft of fencing must be purchased.

Cost of fencing = 150 × £5 = £750

Time taken to complete the job = (146·7 ÷ 8) hours = 18·34 hours

This must be rounded up to 19 hours.

Cost of labour = 19 × £16 = £304

Cost of specialist equipment = £50

To find the cost of erecting the fence around the garden, add together the cost of the fencing, the cost of the labour and the cost of the specialist equipment.

Hence total cost of fence = £750 + £304 + £50 = £1104.

c. The quantities are related by the formula $c = kd^2$ where c is the cost in pounds, d is the diameter in metres and k is a constant.

Hence $4800 = k \times 4^2$

$\Rightarrow 4800 = k \times 16$

$\Rightarrow k = 4800 \div 16$

$\Rightarrow k = 300$

Hence the formula is $c = 300d^2$

$c = 300d^2 = 300 \times 3^2 = 300 \times 9 = 2700$

It would cost £2700 to install the pond.

Therefore the total cost of installing the pond and the fence is £1104 + £2700 = £3804.

 DON'T FORGET

Part (b) contains many elements – Pythagoras' theorem, conversion of metric to imperial units, calculations involving money and rounding. Remember that questions are always marked positively – that is, you receive credit for everything you do correctly.

CASE STUDY 3

AT THE BATHS

EXAMPLE:

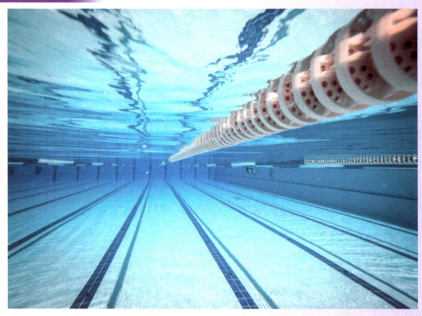

The swimming pool at the local baths is in the shape of a prism.

11 m

The width of the swimming pool is 11 m. The cross-section of the swimming pool is in the shape of a rectangle and a right-angled triangle.

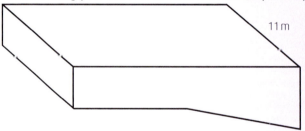

33 m

1·2 m

2·1 m

17 m

(Not drawn to scale)

a. Calculate the volume of the swimming pool. Give your answer in cubic metres.

b. The pool has been emptied for cleaning. Sheila has been told to pump 495 000 litres of water into the empty pool. Calculate the depth of water in the pool at the deep end after this is completed.

c. Water can be pumped into the empty pool at the rate of 420 litres per minute. If the task is required to be completed by 1430 hours on Wednesday, when should she start to fill the pool?

contd

DON'T FORGET

In some questions where there are several parts, the working for part (b) and any further parts may depend on your answer to part (a). If you are unsure about part (a), you can still try the subsequent parts as long as you have an answer. Even if the answer you have to part (a) is incorrect, you will still be able obtain full marks for the subsequent parts.

SOLUTION:

a. **Method 1**

Area of cross-section = area of rectangle + area of triangle

$$= (\text{length} \times \text{width}) + \left(\tfrac{1}{2} \times \text{base} \times \text{height}\right)$$
$$= (33 \times 1{\cdot}2) + [0{\cdot}5 \times (33 - 17) \times (2{\cdot}1 - 1{\cdot}2)]$$
$$= (33 \times 1{\cdot}2) + (0{\cdot}5 \times 16 \times 0{\cdot}9)$$
$$= 39{\cdot}6 + 7{\cdot}2$$
$$= 46{\cdot}8 \, \text{m}^2$$

Volume of prism = area of base × height

$$= 46{\cdot}8 \times 11$$
$$= 514{\cdot}8 \, \text{m}^3$$

Hence the volume of the pool is 514·8 cubic metres.

Method 2

Volume of pool = volume of cuboid + volume of triangular prism

$$= (33 \times 1{\cdot}2 \times 11) + (0{\cdot}5 \times 16 \times 0{\cdot}9 \times 11)$$
$$= 435{\cdot}6 + 79{\cdot}2$$
$$= 514{\cdot}8 \, \text{m}^3$$

Hence the volume of the pool is 514·8 cubic metres.

b. As 1 m = 100 cm, then 1 m³ = (100 × 100 × 100) cm³

$$= 1\,000\,000 \, \text{cm}^3$$
$$= 1\,000 \, \text{litres}$$

495 000 litres = (495 000 ÷ 1000) m³ = 495 m³

Hence area of cross-section = (495 ÷ 11) m² = 45 m².

Method 1

Let height at shallow end = h metres

Area of cross-section = $(33 \times h) + (0{\cdot}5 \times 16 \times 0{\cdot}9)$

$$= (33 \times h) + 7{\cdot}2$$

Hence

45 = $(33 \times h) + 7{\cdot}2$

Leading to

$h = (45 - 7{\cdot}2) \div 33 = 1{\cdot}15$ (correct to two decimal places)

Hence the height at the deep end = (1·15 + 0·9) m = 2·05 m.

Method 2

Volume of water in pool = volume of cuboid + volume of triangular prism

$$= (33 \times h \times 11) + (0{\cdot}5 \times 16 \times 0{\cdot}9 \times 11)$$

Hence

495 = $(363 \times h) + 79{\cdot}2$

Leading to

$h = (495 - 79{\cdot}2) \div 363 = 1{\cdot}15$ (correct to two decimal places)

Hence the height at the deep end = (1·15 + 0·9) m = 2·05 m.

c. Time taken to pump water into pool = (495 000 ÷ 420) minutes

$$= 1178{\cdot}5714 \, \text{minutes}$$
$$= (1178{\cdot}5714 \div 60) \, \text{hours}$$
$$= 19{\cdot}6429 \, \text{hours}$$
$$= 19 \text{ hours } 39 \text{ minutes}$$

Therefore the task should be started 19 hours 39 minutes before 1430 on Wednesday – that is, at 1851 on Tuesday.

CASE STUDY 4

PAINTING THE ROOM

In the section on Composite shapes and their areas 2, we looked at a case study involving painting and wallpapering a room. The topic of painting and decorating provides a useful case study as it involves areas and composite shapes, taking factors such as doors and windows into account, as well as pricing the whole process.

There are numerous online calculators that can help people with the quantities and costs involved in home decorating. This section considers how one of these online calculators works. Before you can enter data into an online calculator, you have to make several measurements and decisions in your home – for example, the height and length of the walls to be painted, the type of paint to be used and the number of coats of paint required.

CASE STUDY

John and Maria are planning to paint the walls in their sitting room.

They decide to use a paint calculator they found on the internet. Before they can use the online calculator, they need to know the following information:

- the height of the wall above the skirting board
- the sum of the lengths of all the walls to be painted
- the type of paint they will be using and its cost per litre
- the number of coats of paint to be used.

Maria enters the following information about the room into the online calculator:

Room wall area

Wall height from skirting board	2·7 m
Total length of all walls to be painted	18·4 m

Window area

Window height	1·1 m
Window width	1·8 m

Door Area

Door height	2·0 m
Door width	0·9 m

Paint

Type of paint	Non-drip emulsion
Number of coats of paint	2
Cost of paint per litre	£6·79

When Maria presses the calculate button, the paint calculator gives the results shown in the table.

Quantity of paint required	7·65 l
Total cost	£51·94

contd

EXAMPLE:

a. Maria decides to check these results. She calculates the total area, in square metres, to be covered by the paint. Find this area showing all the calculations required.

b. Calculate how many square metres of paint are covered by 1 litre of paint, according to the paint calculator.

c. Explain how the cost of £51·94 was calculated.

d. Why will John and Maria probably have to spend more than £51·94 for the paint they require?

e. John and Maria decide to paint the walls in their study as well. The wall height from the skirting board is 2·4 m and the total length of all walls to be painted is 12·4 m. The window measures 1·4 ×1·2 m and the door measures 2·0 × 0·9 m. They will use two coats of the same type of paint as in their sitting room. If they use the paint calculator, what total cost will appear?

SOLUTION:

a. Room wall area = 2·7 × 18·4 = 49·68 m²

Area of window = 1·1 × 1·8 = 1·98 m²

Area of door = 2·0 × 0·9 = 1·8 m²

Area of wall to be painted = 49·68 − 1·98 − 1·8 = 45·9 m²

Number of coats = 2

Total area to be painted = (2 × 45·9) m² = 91·8 m².

b. The calculator indicates that 7·65 litres are required.

Number of litres	Area m²
7·65	91·8
1	$\frac{1}{7\cdot65} \times 91\cdot8 = 12$

Therefore 1 litre of paint covers 12 m².

c. 1 litre of paint costs £6·79.

Hence the cost of 7·65 litres of paint = 7·65 × £6·79 = £51·94.

d. It will be impossible to buy a tin of paint containing exactly 7·65 litres of paint. Therefore John and Maria will have to buy more than 7·65 litres of paint and this will cost more than £51·94.

e. Room wall area = 2·4 × 12·4 = 29·76 m²

Area of window = 1·4 × 1·2 = 1·68 m²

Area of door = 2·0 × 0·9 = 1·8 m²

Area of wall to be painted = 29·76 − 1·68 − 1·8 = 26·28 m²

Number of coats = 2

Total area to be painted = (2 × 26·28) m² = 52·56 m²

Number of litres required = 52·56 ÷ 12 = 4·38

Total cost = 4·38 × £6·79 = £29·74.

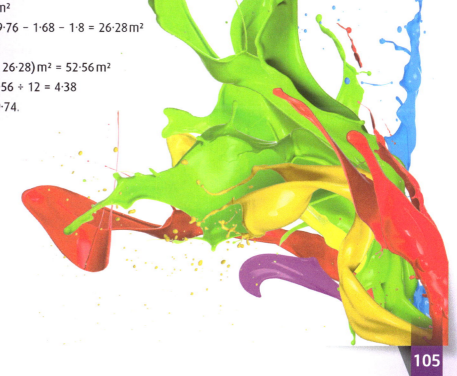

THINGS TO DO AND THINK ABOUT: CASE STUDIES

This section presents a selection of medium to long case studies of the type which could appear in an assessment. You may use a calculator for all the questions.

EXAMPLE:

1. The following information was taken from an advertisement for performances of Bizet's opera *Carmen* at the Theatre Royal:

Performance dates:	Saturday 6 February to Saturday 20 February	
	No Sunday performance	
Performance times:	Evenings	7:15 pm
	Afternoons (Wednesday and Saturday only)	2:30 pm

   a. On how many days was the opera performed?

   b. How many performances of *Carmen* were given altogether?

   c. The running time of the opera is 2 hours 30 minutes, with two intervals of 20 minutes each. At what time does the evening performance end?

   d. Tickets for the afternoon performance cost £24 each, with one free ticket given for every ten purchased.

   Calculate the total cost of tickets for a party of 25 pupils attending this performance.

2. Giovanna works on the production line at a shipbuilding firm. She is paid £9·80 per hour for a 40-hour week.

   The following overtime rates apply:

   - Monday–Friday: time and a quarter
   - Saturday (am): time and a half
   - Saturday (pm) and Sunday: double time.

   In addition to the basic 40 hours, Giovanna works overtime in one week as shown in the table.

Day	Time on	Time off
Friday	1730	2030
Saturday	0830	1400
Sunday	0845	1200

   a. Calculate the number of hours in the week for which overtime is paid at:
      i. time and a quarter
      ii. time and a half
      iii. double time.

   b. What is Giovanna's gross wage for the week?

3. A box in the shape of a cylinder has a height of 30 cm and a diameter of 7·5 cm.

   a. Calculate the volume of the box. Give your answer correct to three significant figures.

   The box is used to contain tennis balls of diameter 7·5 cm. The balls are in contact with each other and with the sides of the cylinder.

   b. Calculate the volume, in cubic centimetres, of the air surrounding the tennis balls once they are placed inside the box.

   c. The tennis balls are to be vacuum-sealed by removing 70% of the air surrounding them. Calculate the volume of air that has to be removed from the box.

contd

4. A supermarket sells two different packs of absorbent paper kitchen towels.

The following information appears on the packs:

PACK A	PACK B
Three rolls of towels	Three rolls of towels
Average 120 sheets per roll	Average 80 sheets per roll
Sheet size 225 × 250 mm	Sheet size 242 × 268 mm
Price £5·80	Price £3·00

a. Calculate the total area of pack A in square metres.

b. Calculate the cost per square metre for pack A.

c. Which pack offers the better value? Justify your answer.

5. The stem and leaf diagram shows the heights to the nearest centimetre of a group of girls.

```
15 │ 0   2   6
16 │ 1   1   2   4   4   5   6   8
17 │ 0   2   2   4   8
18 │ 4
```

(n = 17)   15 │ 10 represents 150 cm

a. Using this information, find.

   i. the median;

   ii. the lower quartile;

   iii. the upper quartile.

b. Draw a boxplot to illustrate these data.

   A sample of boys from the same course was taken. The heights of these students to the nearest centimetre were recorded. The boxplot illustrates the new data.

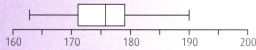

```
   160        170        180        190        200
```

c. By comparing the boxplots, make two appropriate comments about the heights of the boys and girls.

6. A survey was carried out to find the number of books carried by each girl in a group of schoolchildren. The results are shown in the following frequency table.

Number of books	Tally marks	Frequency
0	I	1
1	III	3
2	IIII	5
3	HHt	5
4	HHt I	6
5	III	3
6	II	2

a. From these data, find:

   i.   the number of girls surveyed;

   ii. the median;

   iii. the lower quartile;

   iv. the upper quartile;

   v.   the interquartile range.

b. In the same survey, the number of books carried by each boy was also recorded. The interquartile range was 1·5. Make an appropriate comment comparing the distribution of data for the girls and the boys.

THINGS TO DO AND THINK ABOUT: CASE STUDIES (contd)

7. Nicola has a store card for Neptune Home Improvements.

Neptune Home Improvements
• 15% off your first purchase
• Up to £500 instant credit
• 2·25% monthly interest rate
• Minimum repayments of 5% of balance

Interest is charged on any outstanding balance at the end of the month.

a. Nicola's first purchase using her store card is a fitted kitchen for £2780. How much does she save on this purchase?

b. A week later Nicola buys a dinner set for £63·20 in the store and charges it to her store card. How much does she owe now?

c. Nicola's statement arrives at the end of the month. What is the balance on her store card?

d. Nicola makes the minimum payment. What is her new balance?

e. Calculate the annual percentage rate (APR) for the store card.

8. Joseph works on the production line in a factory. Joseph's November payslip is only partly completed:

Name J. O'Donnell	Employee No. 014	Tax code 1060L	Month November
Basic pay £1584	**Overtime pay**	**Bonus** –	**Gross pay**
Nat. insurance £161·75	**Income tax** £227·25	**Pension**	**Deductions**
			Net pay

a. Joseph worked 160 hours to earn his basic pay during November. Calculate his basic hourly rate of pay.

b. During November, Joseph worked 16 hours overtime at time and a half and 10 hours overtime at double time. Each month he pays 6% of his gross pay into his pension. Calculate his net pay for the month.

c. Joseph pays the following bills from his bank account each month.

He puts aside £400 monthly for other items such as food, transport and entertainment each month and saves the remainder. How much did he save in November?

Electricity	£45
Gas	£24
Telephone	£43
Council tax	£145
Mortgage	£204·16

d. Joseph is hoping to start his own business. He decides to use £500 each month from his savings to take out a loan to fund this enterprise. The following table shows the payments to be made, with payment protection, when £1000 is borrowed from the Rhino Loan Company.

Payments with payment protection.				
APR (%) on £1000	12 months	24 months	36 months	48 months
10	£96·22	£53·54	£39·23	£31·93
12	£97·76	£54·47	£40·20	£32·94
14	£98·66	£55·39	£41·17	£33·95
16	£99·54	£56·30	£42·14	£34·97

Joseph decides to borrow from the Rhino Loan Company. He wants the biggest loan he can get over 36 months with payment protection. Calculate how much he can borrow at 10% APR. Give your answer to the nearest £100.

THINGS TO DO AND THINK ABOUT: EXTRA SECTIONS ANSWERS

MANAGING FINANCE

Planning a budget

1. **a.** She will have a surplus because £821 < £1056·50
    b. She will still have a surplus because £1031 < £1056·50
2. 7 months

Earning money 1

£659·93

Earning money 2

1. £4870
2. £535
3. £439·20

Income tax and other deductions

1. £1030
2. £13 513
3. £29 403

Payslips

A = £2581·25, B = £723·61, C = £229·15

Foreign exchange

a. 655 320 HUF
b. €662·66
c. £79·90

Spending money

a. The 24-pack is best value (£0·40 per bar).
b. Three small packs will cost £5·76 in the offer. This works out at £0·32 per bar. This is now the best value as £0·32 < £0·40.

Finding the best deal

1. The 4 litre bottle is best value as it costs 29·75p per litre, which is less than the other two bottles (33p and 30·67p per litre).
2. **a.** Coverall appears to be best value at £5·50 per litre. This is less than the other paints, which cost £5·70 and £6·10 per litre.
    b. Epaint is the best value because it costs £0·36 per square metre of coverage. This is less than the other paints, which cost £0·39 and £0·38 per square metre.

Interest

1. £131·25
2. **a.** Monthly payments = £825 (12 months), £450 (24 months), £325 (36 months).
    b. £300

Borrowing money

1. £2257·64
2. £5400
3. £196·80

Credit cards

1. **a.** 6279, 6297, 6729, 6792, 6927, 6972
    b. $\frac{3}{4}$
2. A = £212·70; B = £10·64

Probability

1. **a.** $\frac{1}{20}$
    b. $\frac{1}{100}$

Pie charts

1. 75 guests chose chicken.
2. See pie chart:

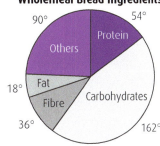

Wholemeal Bread Ingredients

Comparing data sets

Mean = 53; mode = 48; median = 50; range = 32

The interquartile range and boxplots

a.

b. 5·5

Standard deviation

Mean = 20; standard deviation = 3·03

Scattergraphs

a./b.

Taxi Fares

c. 39 miles (approximately)

contd

THINGS TO DO AND THINK ABOUT: EXTRA SECTIONS ANSWERS (contd)

GEOMETRY AND MEASURES

Related quantities 1

1. **a.** 25 days
   b. Friday 8 April 2016
   c. 60 person-days
   d. 5 days

Related quantities 2

1. £42·60
2. **a.** $W = 12·5L$
   b. 325 g

Scale drawings

1256 m²

Navigation

1. **a./b.**

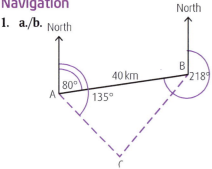

2. **a.**

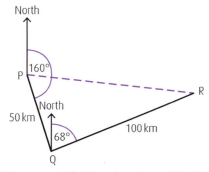

   b. Distance = 110±5 km; bearing = 275±5°

Container packaging

1. 24
2. 160
3. If he packs the boxes so that the 32 cm sides lie along the 3·5 m length, he can fit 240 boxes in the van (10 × 8 × 3). If he packs the boxes so that the 25 cm sides lie along the 3·5 m length, he can fit 252 boxes in the van (14 × 6 × 3).
   Therefore the maximum number is 252.

Precedence tables

Order of activities	Activities that could be done at the same time
C	
G	
F	
D	B
H	
A	
E	

Time management

1955

Tolerance

a. 20°C
b. 22 ± 2°C means that 20–24°C is a suitable range for best growth. Therefore the conditions in the greenhouse are likely to result in best growth.

Gradient

1. No, because 0·0666 > 0·05.
2. $\frac{1}{2}$

Composite shapes 1

a. 10 346 m²
b. 25 laps

Composite shapes 2

a. 174 m²
b. £3500

Volume 1

1. 11 000 cm³
2. 324 cm³
3. 2410 cm³
4. 2140 cm³
5. 7·96 cm

Volume 2

0·32 m³

The theorem of Pythagoras 1

Yes, because the diagonal of the tray is 47·2 cm, which is less than 50 cm.

The theorem of Pythagoras 2

1. 20·2 cm
2. $d = 1·20$ m

contd

NUMERACY

Rounding

1. 6·009
2. 6·0
3. 17
4. 6790
5. 570 cm²

Decimals

1. **a.** 25 400;
   b. 15·3
2. £181
3. **a.** 71·67;
   b. Yes (71·67 < 72)

Fractions

1. Lily, because $\frac{3}{10}\left(=\frac{6}{20}\right) > \frac{1}{4}\left(=\frac{5}{20}\right)$
2. £294
3. £12·60

Percentages

1. 26%
2. No. His marks were 75% for Chemistry, 76% for Physics and 72% for Biology, therefore Physics was his best subject.

Appreciation

1. 2040
2. £526·71
3. **a.** 12%
   b. £275 000

Depreciation

£7900

Ratios

Robert receives £250, Catriona £200 and Louis £150

Distance, speed and time

a. 2 hours 30 minutes
b. 50 mph

Reading from a scale

900 ml

Stem and leaf diagrams

a.

Rovers   United

```
              9 | 1 | 7 8 9
  9 8 7 6 5 4 3 3 2 | 2 | 0 1 2 2 3 4 4 5 5 6 6 8 9
      4 3 3 2 1 0 0 | 3 | 2 4
                  0 | 4 |
```

n = 18   n = 18

```
                    | 1 |          7        means 17
                  9 | 1 |                   means 19
```

b. The United squad is younger than the Rovers squad.

Line graphs

Costa del Sol – Temperature

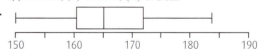

Information from tables and graphs 1

£60·20

Information from tables and graphs 2

a 27·46
b. She should lose around 14·3 pounds.

Maps and plans

Approximately 50 km.

Things to do and think about: case studies

1. **a.** 13 days
   b. 18 performances
   c. 10.25 pm
   d. £552
2. **a.** (i) 3 hours; (ii) 3 hours 30 minutes;
       (iii) 5 hours 15 minutes
   b. £583·10
3. **a.** 1330 cm³
   b. 442 cm³ (to three significant figures)
   c. 309 cm³ (to three significant figures)
4. **a.** 20·25 m²
   b. 28·6 pence
   c. Pack B costs 19·3p per m² and is better value because this is less than 28·6p per m².
5. **a.** (i) 165 cm; (ii) 161 cm; (iii) 172 cm
   b.

   ```
   |----| _____ |-------------------|
   150      160      170      180      190
   ```

   c. The boys were taller and the heights of the girls were more spread out.
6. **a.** (i) 25; (ii) 3; (iii) 2; (iv) 4; (v) 2
   b. The number of books carried by the boys was more consistent.
7. **a.** £417
   b. £2426·20
   c. £2480·79
   d. £2356·75
   e. 30·6%
8. **a.** £9·90
   b. £1509·42
   c. £648·26
   d. £12 700

GLOSSARY OF MATHEMATICAL TERMS

Angle of elevation: the angle between the horizontal and the line of sight from an observer to an object.

Annual percentage rate (APR): the annual rate charged for borrowing money, expressed as a single percentage number, representing the yearly cost of the loan

Best-fitting line: drawn on a scattergraph to follow the pattern of points

Bonus: an extra payment given for meeting targets, for producing exceptional work, or for completing a job ahead of schedule

Budget: a way of planning income and expenditure to ensure you are only spending what you can afford

Case study: a longer question, usually in several parts, which focuses on a particular theme – for example, decorating a room

Circumference: the distance around the outside of a circle found using the formula $C = \pi d$

Commission: a way of rewarding salespersons by giving them a percentage of their sales as part of their salary

Composite shape: a shape made up of two or more shapes – for example, a cylinder surmounted by a hemisphere

Compound bar graph: an extension of an ordinary bar graph that compares two or more quantities at the same time

Correlation: tells us how variables such as height and weight are related

Credit card statement: a monthly statement sent to all cardholders showing all recent transactions, the amount owed (the balance) and the minimum amount that must be paid back

Critical path: the longest path through a network of activities; it gives the minimum time required to complete a job

Deficit: this occurs when you spend (expenditure) more than you can afford (income)

Deductions: money taken off your gross pay; the main deductions are income tax and National Insurance

Direct proportion: two quantities are said to be in direct proportion if, as one quantity increases, the other increases at the same rate

Dividend: a sum of money paid annually to its shareholders by a company

Extrapolation: the process of estimating the value of a variable that lies outside a known range of values on a line graph – often used to predict future events

Expected frequency: the number of times you would expect a particular event to happen in an experiment

Five-figure summary: this is a summary of a numerical data set consisting of the lowest value, the lower quartile, the median, the upper quartile and the highest value.

Gradient: the slope of a straight line

Gross pay: income before any deductions are taken off

Hire purchase: a way of buying expensive items by paying a small part of the cost, called the deposit, followed by monthly instalments

Hypotenuse: the side opposite the right angle in a right-angled triangle

Indirect (or inverse) proportion: two quantities are said to be in indirect proportion if, as one quantity increases, the other decreases at the same rate

Interpolation: the process of estimating the value of a variable at a point in between values on a line graph

Interquartile range: a measure of spread given by $Q_3 - Q_1$

Justify: explain the solution to a problem in words

Least common multiple: the smallest multiple common to two or more numbers

Lower quartile: denoted by Q_1; the smallest of the three quartiles that divide a data set into four equal parts

Mean, mode, median: three measures of central tendency used to analyse data sets

Mortgage: a legal agreement that allows you to take out a loan from a bank or building society to buy a house

Multiplier: the number we multiply by to carry out a percentage increase or decrease

Net pay: pay after deductions have been taken off your gross pay; it is also called take-home pay

Overtime: a way of earning more money by working beyond normal working hours; usually paid at rates of time and a half or double time

Perimeter: the distance around the outside of a shape

Perpendicular bisector: the perpendicular bisector of a line is a straight line which crosses the midpoint of the line at right angles.

Personal allowance: the amount of a person's income that is not subject to income tax

Piecework: payment for each item (or piece) a worker makes

Precedence table: a table that can be used to plan activities based on which activities must follow others and which activities can be carried out at the same time

Probability: a measure of how likely an event is to happen

Ratio: a ratio compares two values, e.g. there are 3 girls for every 2 boys in a classroom.

Ready reckoner: a table of numbers used to ease calculations; often used to convert units – for example, metric to imperial units

Related quantities: quantities such as distance and time that are related by direct or indirect proportion

Representative fraction: shows the scale on a map – for example, 1:1000; independent of units

Scattergraph: a statistical diagram used to compare two data sets

Sector: an area bounded by two radii and an arc

Semi-interquartile range: a measure of spread given by $\frac{1}{2}(Q_3 - Q_1)$

Space diagonal: in a cuboid, a space diagonal is a line that goes from a vertex of the cuboid, through the centre of the cuboid, to the opposite vertex.

Standard deviation: a measure of spread given by
$$s = \sqrt{\frac{\Sigma(x - \overline{x})^2}{n-1}} = \sqrt{\frac{\Sigma x^2 - (\Sigma x)^2/n}{n-1}},$$
where n is the sample size

Stem and leaf diagram: an ordered statistical diagram in which each member of the data set is split into a 'stem' and a 'leaf'

Surplus: this occurs when you spend (expenditure) less than you earn (income)

The theorem of Pythagoras: in a right-angled triangle, the square of the hypotenuse is equal to the sum of the squares on the other two sides; usually given by the formula $a^2 + b^2 = c^2$

Three-figure bearings: describe directions relative to north (000°); the three-figure bearings of other directions are given by angles measured clockwise from north

Tolerance: the range of values in which a measurement is acceptable – for example $36 \pm 0.5\,\text{cm}$

Upper quartile: denoted by Q_3; the largest of the three quartiles that divide a data set into four equal parts